Number Theory

www.lulu.com
Lulu Press, Inc
627 Davis Drive, Suite 300,
Morrisville, NC 27560.

Author Affiliations

Ms. Manjula R.,
Assistant Professor,
S&H department,
Vignan's Institute of Management and Technology for Women,
Kondapur, Medchal, Malkajgiri, Hyderabad, Telangana – 501301.

Ms. Varalaxmi K.,
Assistant professor,
S&H department,
Nalla Narasimha Reddy Group of Institutions,
Narapally, Ghatkesar, Hyderabad, Telangana – 500088.

Mr. Suresh T.,
Assistant Professor & Head of the Faculty- Maths
Bhoj Reddy Engineering College for Women,
Vinay Nagar, Saidabad, Hyderabad – 500059.

Ms. Rajitha J.,
Assistant Professor,
S&H department
Vignan's Institute of Management and Technology for Women,
Kondapur, Medchal, Malkajgiri, Hyderabad, Telangana – 501301.

First Printing: 2023

ISBN: 978-1-312-34689-5

Copyright License @ Manjula R., Varalaxmi K., Suresh T., Rajitha J.

This book has been published with all reasonable efforts to make the material error-free after the author's consent. No part of this book shall be used, or reproduced in any manner, without the author's permission, except for brief quotations embodied in critical articles and reviews.

The Author of this book is solely responsible and liable for its content, including but not limited to the views, representations, descriptions, statements, information, opinions, and references ["Content"]. The Content of this book shall not constitute or be construed or deemed to reflect the opinion or expression of the Publisher or Editor. Neither the Publisher nor Editor endorse or approve the Content of this book or guarantee the reliability, accuracy, or completeness of the Content published herein and do not make any representations or warranties of any kind, express or implied, including but not limited to the implied warranties of merchantability, fitness for a particular purpose. The Publisher and Editor shall not be liable whatsoever for any errors or omissions, whether such errors or omissions result from negligence, accident, or any other cause or claims for loss or damages of any kind, including without limitation, indirect or consequential loss or damage arising out of use, inability to use, or about the reliability, accuracy or sufficiency of the information contained in this book. This book was written based on Intelligence with the support of various sources.

Number Theory

By

Ms. Manjula R.

Ms. Varalaxmi K.

Mr. Suresh T.

Ms. Rajitha J.

2023

About the Authors

Ms. Manjula R., M.Sc (Mathematics), is an Assistant Professor with 14 years of teaching experience. She authored 4 paper publications and presented and participated in various national and international conferences.

Ms. Varalaxmi K., M.Sc (Mathematics), is an Assistant Professor with 15 years of teaching experience. She authored 4 paper publications and presented and participated in various national and international conferences. She also has a membership in ISTE.

Ms. Suresh T., M.Sc (Mathematics) & B.Ed, is an Assistant Professor and Head (Mathematics) with 16 years of teaching experience. He authored a few paper publications and presented and participated in various national and international conferences.

Ms. Rajitha J., M.Sc (Mathematics), is an Assistant Professor with 14 years of teaching experience. She authored a few paper publications and presented and participated in various national and international conferences.

About Book

This Book of Number Theory is a captivating exploration of one of the oldest branches of mathematics. From its ancient origins to modern-day breakthroughs, this book uncovers the patterns, properties, and relationships that lie at the heart of numbers. It delves into topics such as divisibility, prime numbers, modular arithmetic, Diophantine equations, prime number distribution, and sieve methods. With its comprehensive coverage and engaging explanations, The Book of Number Theory reveals the beauty and significance of this fascinating mathematical discipline.

Table of Contents

1. Introduction to Number Theory 1
    1.1 History and Significance 3
    1.2 Fundamental Concepts 5
2. Divisibility and Factors 8
    2.1 Divisibility Rules 9
    2.2 Prime Factorization 10
    2.3 Greatest Common Divisor (GCD) 12
    2.4 Least Common Multiple (LCM) 14
3. Prime Numbers 16
    3.1 Sieve of Eratosthenes 17
    3.2 Prime Number Properties 19
    3.3 Prime Number Distribution 21
    3.4 Prime number generation algorithms 23
4. Modular Arithmetic 25
    4.1 Congruence relation 26
    4.2 Modular addition, subtraction, and multiplication 28
    4.3 Modular Exponentiation 29
    4.4 Modular Inverses 31
5. Diophantine Equations 34
    5.1 Linear Diophantine Equations 35
    5.2 Pythagorean Triples 37
    5.3 Fermat's Last Theorem 39
6. Euler's Totient Function 41
    6.1 Definition and Properties 42

- 6.2 Euler's Theorem 44
- 6.3 Applications in Cryptography 45
- 7. Quadratic Residues 48
  - 7.1 Quadratic residues and non-residues 49
  - 7.2 Quadratic Reciprocity Theorem 51
  - 7.3 Legendre and Jacobi Symbols 52
- 8. Continued Fractions 55
  - 8.1 Definition, Properties, and Applications 55
  - 8.2 Convergents and Approximations 57
  - 8.3 Pell's Equation 59
- 9. Cryptography 61
  - 9.1 RSA Encryption Algorithm 63
  - 9.2 Diffie-Hellman Key Exchange 65
  - 9.3 Elliptic Curve Cryptography 67
- 10. Number-Theoretic Functions 70
  - 10.1 Euler's phi Function 72
  - 10.2 Mobius Function 73
  - 10.3 Riemann zeta Function 75
- 11. Prime Number Theorems 78
  - 11.1 Prime Number Theorem 79
  - 11.2 Distribution of Prime Numbers 80
  - 11.3 Riemann Hypothesis 82
- 12. Unsolved Problems in Number Theory 85
  - 12.1 Goldbach's Conjecture 87
  - 12.2 Twin Prime Conjecture 88
  - 12.3 Collatz Conjecture 89

13. Conclusion and Further Exploration 91

Number Theory

1. Introduction to Number Theory

Number theory is a branch of mathematics that deals with the properties and relationships of numbers, primarily integers. It explores patterns, structures, and properties within the realm of numbers and seeks to understand their fundamental nature. Number theory has a long history and has played a significant role in both pure mathematics and practical applications.

Number theory encompasses various fundamental concepts, including prime numbers, divisibility, modular arithmetic, Diophantine equations, and congruences. These concepts form the building blocks for exploring more advanced topics within the field.

The study of prime numbers is a central aspect of number theory. Prime numbers are positive integers greater than 1 that have no divisors other than 1 and themselves. Understanding the distribution, properties, and factorization of prime numbers is of fundamental importance in number theory.

Divisibility is another fundamental concept in number theory. It involves examining the relationship between two numbers and whether one number divides evenly into the other without leaving a remainder. Divisibility rules and properties play a crucial role in number theory, providing insights into prime factorization and other aspects of number theory.

Modular arithmetic is a system of arithmetic that deals with numbers' remainders when divided by a fixed positive integer called the modulus. It focuses on operations such as addition, subtraction, multiplication, and exponentiation performed on these remainders. Modular arithmetic has applications in cryptography, computer science, and various other areas of mathematics.

Number Theory

Diophantine equations are polynomial equations where the solutions are sought in the integers. These equations, named after the ancient Greek mathematician Diophantus, are particularly important in number theory. Solving Diophantine equations involves finding integer solutions that satisfy the given equation, often leading to intriguing patterns and properties.

Congruences are a concept in number theory that deals with equivalence classes of numbers based on their remainders when divided by a fixed modulus. Two numbers are said to be congruent modulo a given modulus if they leave the same remainder when divided by that modulus. Congruences have wide-ranging applications, including modular arithmetic, solving equations, and analyzing number patterns.

Number theory has significance beyond its theoretical exploration. It finds practical applications in areas such as cryptography, coding theory, computer science, and information security. Cryptography relies on number-theoretic principles to develop secure encryption algorithms, while coding theory utilizes number theory to design error-correcting codes for reliable data transmission and storage.

In conclusion, number theory is a branch of mathematics that focuses on the properties and relationships of numbers. It has a rich history and encompasses fundamental concepts such as prime numbers, divisibility, modular arithmetic, Diophantine equations, and congruences. Number theory plays a crucial role in pure mathematics, serving as a foundation for various branches, and finds practical applications in cryptography, computer science, and other fields.

1.1 History and Significance

History of Number Theory:

Number theory has a long and rich history, dating back to ancient civilizations. The Egyptians, Babylonians, and ancient Greeks were among the earliest civilizations to explore the properties of numbers. However, the formal development of number theory as a mathematical discipline began in ancient Greece with mathematicians such as Pythagoras, Euclid, and Diophantus.

Pythagoras and his followers, known as the Pythagoreans, believed in the mystical and symbolic significance of numbers. They discovered important properties of numbers, particularly whole numbers, and ratios, which led to the development of the Pythagorean theorem and other mathematical principles.

Euclid, in his monumental work "Elements," presented a systematic approach to geometry and number theory. He established the foundations of number theory by proving theorems related to prime numbers, divisibility, and number patterns. His work served as a basis for further exploration and development in number theory.

During the Middle Ages, Islamic mathematicians made significant contributions to number theory. Scholars such as Al-Khwarizmi and Omar Khayyam worked on topics like solving quadratic equations and understanding the properties of integers and fractions.

In the Renaissance period, number theory received attention from prominent mathematicians like Pierre de Fermat, who introduced the concept of Fermat's Last Theorem, and René Descartes, who developed coordinate geometry and laid the groundwork for algebraic number theory.

In the 18th and 19th centuries, number theory experienced remarkable progress with the contributions of mathematicians like Leonhard Euler, Carl Friedrich Gauss, and Évariste Galois. Gauss made significant advancements in areas such as modular arithmetic, quadratic reciprocity, and the distribution of prime numbers. Galois revolutionized the field with his work on the theory of equations, establishing the foundation for abstract algebra.

Significance of Number Theory:

Number theory has both theoretical and practical significance in mathematics and various other fields:

Pure Mathematics: Number theory is a fundamental branch of pure mathematics. It explores the inherent properties and structures of numbers, offering deep insights into the nature of mathematics itself. It provides a framework for studying patterns, relationships, and properties of numbers, prime numbers, divisibility, and algebraic structures.

Cryptography: Number theory plays a critical role in cryptography, the science of securing communication and data. Cryptographic algorithms rely on number-theoretic principles, such as prime factorization, modular arithmetic, and discrete logarithms, to encrypt and decrypt sensitive information. The security of many encryption schemes is based on the difficulty of certain number-theoretic problems.

Coding Theory: Number theory is essential in coding theory, which deals with error detection and correction in data transmission. Error-correcting codes use number-theoretic concepts to encode and decode information, ensuring reliable communication and storage.

Computer Science: Number theory has applications in computer science and algorithm design. Efficient algorithms for prime testing, factorization, and modular arithmetic are crucial in various computational problems, including number theory algorithms, cryptography, and data structures.

Mathematical Research: Number theory continues to be a vibrant field of research with many unsolved problems. Notable open problems include the Riemann Hypothesis, the Goldbach Conjecture, and the Twin Prime Conjecture. These unsolved problems provide challenges and motivate further exploration and discovery.

In conclusion, number theory has a long and significant history, starting from ancient civilizations to the present day. It is a foundational branch of mathematics, that explores the properties and relationships of numbers. Number theory has practical applications in cryptography, coding theory, and computer science, and plays a crucial role in theoretical research, providing insights into the nature of numbers and mathematics itself.

1.2 Fundamental Concepts

Number theory encompasses several fundamental concepts that form the basis of its study. These concepts are essential for understanding the properties and relationships of numbers. Here are some of the fundamental concepts in number theory:

Prime Numbers: Prime numbers are positive integers greater than 1 that have exactly two positive divisors: 1 and the number itself. Examples of prime numbers include 2, 3, 5, 7, 11, and so on. Prime numbers play a central role in number theory, and their properties and distribution have been studied extensively.

Divisibility: Divisibility is a concept that deals with how one number divides another. A number is said to be divisible by another number if the division results in an integer without leaving a remainder. For example, 15 is divisible by 3 because 15 divided by 3 equals 5 without any remainder. Divisibility rules help determine if a number is divisible by another number without performing the actual division.

Prime Factorization: Prime factorization involves expressing a given number as a product of prime numbers. Every positive integer greater than 1 can be uniquely expressed as a product of prime numbers, known as its prime factorization. For example, the prime factorization of 30 is 2 × 3 × 5. Prime factorization is crucial for various number theory concepts, such as finding common factors, computing the greatest common divisor, and simplifying fractions.

Modular Arithmetic: Modular arithmetic is a system of arithmetic that deals with numbers' remainders when divided by a fixed positive integer called the modulus. It focuses on operations such as addition, subtraction, multiplication, and exponentiation performed on these remainders. Modular arithmetic has applications in cryptography, number theory algorithms, and other fields.

Diophantine Equations: Diophantine equations are polynomial equations where the solutions are sought in the integers. The study of Diophantine equations involves finding integer solutions that satisfy the given equation. Famous examples include Pythagorean triples, which satisfy the equation $a^2 + b^2 = c^2$, where a, b, and c are integers.

Congruences: Congruences are a concept in number theory that deals with equivalence relations between numbers based on their remainders when divided by a fixed modulus. Two numbers are said to be congruent modulo a given modulus if they leave the same remainder when divided by that modulus. Congruences have applications in solving equations, proving divisibility properties, and exploring number patterns.

These fundamental concepts provide a strong foundation for exploring more advanced topics in number theory, such as quadratic residues, number-theoretic functions, continued fractions, and prime number distribution. They are essential tools for understanding the properties and relationships of numbers and form the basis for further investigations in the field of number theory.

2. Divisibility and Factors

Divisibility:

Divisibility is a fundamental concept in number theory that deals with how one number divides another. A number is said to be divisible by another number if the division results in an integer without leaving a remainder. The following terms and concepts are related to divisibility:

Divisor: A divisor of a given number divides that number without leaving a remainder. For example, 2 and 3 are divisors of 6 because 6 divided by 2 and 6 divided by 3 both result in integers (3 and 2, respectively).

Multiple: A multiple of a given number is obtained by multiplying it by any positive integer. For example, multiples of 4 include 4, 8, 12, 16, and so on.

Prime Numbers: Prime numbers are positive integers greater than 1 that have exactly two distinct positive divisors: 1 and the number itself. Prime numbers cannot be further divided into smaller factors. Examples of prime numbers include 2, 3, 5, 7, 11, and so on.

Composite Numbers: Composite numbers are positive integers greater than 1 that are not prime. Composite numbers have more than two positive divisors because they can be factored into smaller prime numbers. Examples of composite numbers include 4, 6, 8, 9, 10, and so on.

Prime Factorization: Prime factorization involves expressing a given number as a product of prime numbers. Every positive integer greater than 1 can be uniquely expressed as a product of prime factors. For example, the prime factorization of 30 is $2 \times 3 \times 5$.

Greatest Common Divisor (GCD): The greatest common divisor of two or more numbers is the largest positive integer that divides all the given numbers without leaving a remainder. GCD is often computed using the prime factorization method.

Factors:

Factors are closely related to divisibility. A factor of a given number is a number that divides it without leaving a remainder. For example, factors of 12 include 1, 2, 3, 4, 6, and 12. Factors can be classified as follows:

Proper Factors: Proper factors of a number are the factors excluding the number itself. For example, the proper factors of 12 are 1, 2, 3, 4, and 6.

Perfect Numbers: Perfect numbers are positive integers that are equal to the sum of their proper divisors. For example, 6 is a perfect number because its proper divisors (1, 2, 3) add up to 6.

Understanding divisibility and factors is essential for various number theory applications, including prime factorization, determining whether a number is prime or composite, finding common divisors, simplifying fractions, and solving equations involving integers. These concepts serve as building blocks for more advanced topics in number theory and have practical applications in fields such as cryptography, computer science, and algebraic manipulation.

2.1 Divisibility Rules

Divisibility rules are a set of guidelines that help determine if a number is divisible by another number without performing the actual division. These rules provide a quick and efficient way to check for divisibility. Here are some common divisibility rules:

Divisibility by 2: A number is divisible by 2 if its last digit is even (0, 2, 4, 6, or 8). For example, 16, 358, and 4,720 are divisible by 2.

Divisibility by 3: A number is divisible by 3 if the sum of its digits is divisible by 3. For example, 123 (1 + 2 + 3 = 6) and 1,881 (1 + 8 + 8 + 1 = 18) are divisible by 3.

Divisibility by 4: A number is divisible by 4 if the last two digits of the number form a multiple of 4. For example, 1,256 (56 is a multiple of 4) and 284 (84 is a multiple of 4) are divisible by 4.

Divisibility by 5: A number is divisible by 5 if its last digit is 0 or 5. For example, 105 and 3,230 are divisible by 5.

Divisibility by 6: A number is divisible by 6 if it is divisible by both 2 and 3. For example, 180 (divisible by 2 and 3) and 1,032 (divisible by 2 and 3) are divisible by 6.

Divisibility by 9: A number is divisible by 9 if the sum of its digits is divisible by 9. For example, 243 (2 + 4 + 3 = 9) and 1,989 (1 + 9 + 8 + 9 = 27) are divisible by 9.

Divisibility by 10: A number is divisible by 10 if its last digit is 0. For example, 120 and 5,670 are divisible by 10.

These rules provide a convenient way to quickly determine the divisibility of numbers. However, it's important to note that these rules are specific to certain divisors and may not cover all possible divisors. For divisors other than those mentioned above, performing the actual division is necessary to determine divisibility.

2.2 Prime Factorization

Prime factorization is the process of expressing a given number as a product of prime numbers. It is a fundamental concept in

Number Theory

number theory and plays a crucial role in various mathematical applications. Here's how to find the prime factorization of a number:

1. Start with the smallest prime number, which is 2. Check if the given number is divisible by 2. If it is divisible, divide the number by 2 and record the factor 2. Repeat this step until the number is no longer divisible by 2.

2. Move on to the next prime number, which is 3. Check if the number is divisible by 3. If it is divisible, divide the number by 3 and record the factor 3. Repeat this step until the number is no longer divisible by 3.

3. Continue this process with successive prime numbers (5, 7, 11, 13, and so on), checking divisibility and recording the factors until the number becomes 1.

4. The product of all the recorded prime factors gives the prime factorization of the original number.

For example, let's find the prime factorization of 60:

Step 1: 60 is divisible by 2. Divide 60 by 2 to get 30. Record factor 2.

Step 2: 30 is divisible by 2. Divide 30 by 2 to get 15. Record factor 2.

Step 3: 15 is not divisible by 2. Move on to the next prime number, 3. 15 is divisible by 3. Divide 15 by 3 to get 5. Record factor 3.

Step 4: 5 is not divisible by 2 or 3. Move on to the next prime number, 5. 5 is divisible by 5. Divide 5 by 5 to get 1. Record factor 5.

The prime factorization of 60 is $2 \times 2 \times 3 \times 5$, or simply $2^2 \times 3 \times 5$.

Prime factorization provides a unique representation of a number as a product of prime factors. It is helpful in various mathematical

Number Theory

computations, including finding common factors, simplifying fractions, and solving equations involving integers. Prime factorization is also used in applications like cryptography and number theory algorithms.

2.3 Greatest Common Divisor (GCD)

The greatest common divisor (GCD) is a fundamental concept in number theory that refers to the largest positive integer that divides two or more given numbers without leaving a remainder. The GCD has various applications in mathematics and is commonly used in simplifying fractions, finding common factors, and solving equations. Here's how to find the GCD of two numbers:

Method 1: Prime Factorization

1. Find the prime factorization of each number.

2. Identify the common prime factors between the two numbers.

3. Multiply these common prime factors together to obtain the GCD.

For example, let's find the GCD of 48 and 60:

Step 1: Prime factorization of 48: $48 = 2 \times 2 \times 2 \times 2 \times 3 = 2^4 \times 3$.

Prime factorization of 60: $60 = 2 \times 2 \times 3 \times 5 = 2^2 \times 3 \times 5$.

Step 2: Common prime factors: The common prime factors between 48 and 60 are 2 and 3.

Step 3: Multiply the common prime factors: $GCD(48, 60) = 2 \times 3 = 6$.

Therefore, the GCD of 48 and 60 is 6.

Method 2: Euclidean Algorithm

1. Start with the two numbers, a and b, for which you want to find the GCD.

2. Divide the larger number (a) by the smaller number (b) and obtain the remainder (r).

3. Replace a with b and b with r.

4. Repeat steps 2 and 3, dividing the new a by the new b and obtaining a new remainder, until the remainder becomes 0.

5. The GCD is the last non-zero remainder obtained in the process.

For example, let's find the GCD of 48 and 60 using the Euclidean Algorithm:

Step 1: Divide 60 by 48: $60 \div 48 = 1$ remainder 12.

Step 2: Replace a with 48 and b with 12.

Step 3: Divide 48 by 12: $48 \div 12 = 4$ remainder 0.

Since the remainder is now 0, the GCD is the last non-zero remainder, which is 12.

Therefore, the GCD of 48 and 60 is 12.

Both methods, prime factorization, and the Euclidean Algorithm, yield the same result. The Euclidean Algorithm is often more efficient for larger numbers as it involves repeated division. The GCD is an important concept in number theory and finds applications in simplifying fractions, finding common factors, and solving equations involving integers.

Number Theory

2.4 Least Common Multiple (LCM)

The least common multiple (LCM) is a fundamental concept in number theory that refers to the smallest positive integer that is divisible by two or more given numbers. The LCM has applications in various mathematical computations, including finding a common denominator for fractions and solving equations involving multiple variables. Here's how to find the LCM of two numbers:

Method 1: Prime Factorization

1. Find the prime factorization of each number.

2. Identify the highest power of each prime factor that appears in either of the factorizations.

3. Multiply these prime factors together to obtain the LCM.

For example, let's find the LCM of 12 and 18:

Step 1: Prime factorization of 12: $12 = 2 \times 2 \times 3 = 2^2 \times 3$.

Prime factorization of 18: $18 = 2 \times 3 \times 3 = 2 \times 3^2$.

Step 2: Highest powers of prime factors: The highest power of 2 is 2^2, and the highest power of 3 is 3^2.

Step 3: Multiply the prime factors: $LCM(12, 18) = 2^2 \times 3^2 = 4 \times 9 = 36$.

Therefore, the LCM of 12 and 18 is 36.

Method 2: GCD and Division

1. Find the greatest common divisor (GCD) of the two numbers.

2. Divide the product of the two numbers by their GCD.

3. The result is the LCM.

For example, let's find the LCM of 12 and 18 using the GCD:

Step 1: Find the GCD of 12 and 18: GCD(12, 18) = 6.

Step 2: Divide the product of the numbers by their GCD: (12 × 18) ÷ 6 = 216 ÷ 6 = 36.

Therefore, the LCM of 12 and 18 is 36.

Both methods, prime factorization and using the GCD, yield the same result. The LCM is an important concept in number theory and finds applications in various mathematical computations. It helps determine a common denominator for fractions and is used to find the smallest number that satisfies multiple conditions.

3. Prime Numbers

Prime numbers are a fundamental concept in number theory. A prime number is a positive integer greater than 1 that has no divisors other than 1 and itself. In other words, a prime number cannot be evenly divided by any other number except 1 and the number itself.

Here are some key properties and facts about prime numbers:

Examples: The first few prime numbers are 2, 3, 5, 7, 11, 13, 17, 19, 23, and so on. There are infinitely many prime numbers.

Prime Factorization: Every positive integer greater than 1 can be uniquely expressed as a product of prime numbers, known as its prime factorization. For example, the prime factorization of 24 is $2 \times 2 \times 2 \times 3$, or simply $2^3 \times 3$.

Sieve of Eratosthenes: The Sieve of Eratosthenes is an ancient algorithm used to find all prime numbers up to a given limit. It involves iteratively crossing out multiples of prime numbers until all non-prime numbers are eliminated, leaving only the primes.

Prime Number Theorem: The Prime Number Theorem, formulated by mathematician Jacques Hadamard and Charles Jean de la Vallée-Poussin independently in 1896, describes the distribution of prime numbers. It states that as the number n tends towards infinity, the density of primes around n follows a logarithmic growth pattern.

Goldbach's Conjecture: Goldbach's Conjecture, proposed by German mathematician Christian Goldbach in 1742, states that every even integer greater than 2 can be expressed as the sum of two prime numbers. This conjecture remains unsolved and is one of the oldest unsolved problems in number theory.

6. Prime Testing: Primality testing refers to determining whether a given number is prime or composite. Various algorithms exist for primality testing, ranging from trial division to more advanced methods like the Miller-Rabin primality test and the AKS primality test.

Prime numbers have applications in various fields, including cryptography, number theory algorithms, and computer science. They play a crucial role in encryption algorithms like RSA, which rely on the difficulty of factoring large composite numbers into their prime factors for security. Prime numbers also feature prominently in number theory research, contributing to the understanding of number patterns and the distribution of primes.

Exploring prime numbers is an active area of research, with ongoing efforts to discover new prime numbers and investigate their properties. The search for large prime numbers has led to collaborative projects like the Great Internet Mersenne Prime Search (GIMPS), where volunteers use their computer processing power to find new prime numbers.

Overall, prime numbers hold a special place in number theory and mathematics, and their properties and distribution continue to intrigue and challenge mathematicians.

3.1 Sieve of Eratosthenes

The Sieve of Eratosthenes is an ancient algorithm used to find all prime numbers up to a given limit. It was developed by the Greek mathematician Eratosthenes around 200 BCE. The sieve method is an efficient way to identify prime numbers within a range and is based on the following steps:

1. Create a list of numbers from 2 to the desired limit.

Number Theory

2. Start with the first number, 2, and mark it as prime.

3. Proceed to the next unmarked number, which is a prime number. Mark it as prime and cross out all of its multiples.

4. Move to the next unmarked number and repeat the process until all the numbers have been processed or crossed out.

Here's a step-by-step explanation of the Sieve of Eratosthenes algorithm:

1. Create a list of numbers from 2 to the desired limit. Assume we want to find all prime numbers up to 30.

   List: 2 3 4 5 6 7 8 9 10 11 12 13 14 15 16 17 18 19 20 21 22 23 24 25 26 27 28 29 30

2. Start with the first unmarked number, which is 2, and mark it as prime.

   List: 2 (P) 3 4 5 6 7 8 9 10 11 12 13 14 15 16 17 18 19 20 21 22 23 24 25 26 27 28 29 30

3. Cross out all multiples of 2 (excluding 2 itself).

   List: 2 (P) 3 4 (X) 5 6 (X) 7 8 (X) 9 (X) 10 (X) 11 12 (X) 13 14 (X) 15 (X) 16 (X) 17 18 (X) 19 20 (X) 21 (X) 22 (X) 23 24 (X) 25 (X) 26 (X) 27 (X) 28 (X) 29 30 (X)

4. Move to the next unmarked number, which is 3, and mark it as prime.

   List: 2 (P) 3 (P) 4 (X) 5 6 (X) 7 8 (X) 9 (X) 10 (X) 11 12 (X) 13 14 (X) 15 (X) 16 (X) 17 18 (X) 19 20 (X) 21 (X) 22 (X) 23 24 (X) 25 (X) 26 (X) 27 (X) 28 (X) 29 30 (X)

Number Theory

5. Cross out all multiples of 3 (excluding 3 itself).

List: 2 (P) 3 (P) 4 (X) 5 6 (X) 7 8 (X) 9 (X) 10 (X) 11 12 (X) 13 14 (X) 15 (X) 16 (X) 17 18 (X) 19 20 (X) 21 (X) 22 (X) 23 24 (X) 25 (X) 26 (X) 27 (X) 28 (X) 29 30 (X)

6. Repeat the process for the remaining unmarked numbers.

After completing the process, the remaining unmarked numbers in the list are the prime numbers.

List: 2 (P) 3 (P) 5 (P) 7 (P) 11 (P) 13 (P) 17 (P) 19 (P) 23 (P) 29 (P)

Thus, the prime numbers up to 30 are 2, 3, 5, 7, 11, 13, 17, 19, 23, and 29.

The Sieve of Eratosthenes algorithm is an efficient method to find prime numbers within a given range. It eliminates the need for performing division tests for every number, making it faster for larger ranges. The sieve method is widely used in number theory and prime number research.

3.2 Prime Number Properties

Prime numbers have several interesting properties and characteristics that make them unique within the realm of integers. Here are some notable properties of prime numbers:

Divisibility: Prime numbers have only two distinct positive divisors: 1 and the number itself. They cannot be evenly divided by any other number. This property distinguishes prime numbers from composite numbers, which have more than two positive divisors.

Prime Factorization: Every positive integer greater than 1 can be uniquely expressed as a product of prime numbers, known as its

Number Theory

prime factorization. Prime numbers are the building blocks of all other numbers. For example, the prime factorization of 24 is $2 \times 2 \times 2 \times 3$, or simply $2^3 \times 3$.

Infinitude: There are infinitely many prime numbers. This fact was proved by the ancient Greek mathematician Euclid around 300 BCE. Euclid's proof shows that if we assume there are finitely many prime numbers, we can construct a new prime number that contradicts that assumption.

Distribution: Prime numbers are not evenly distributed among all integers. They become less frequent as numbers increase, but there is no predictable pattern to their distribution. Prime numbers exhibit clustering, irregularity, and apparent randomness.

Twin Primes: Twin primes are pairs of prime numbers that differ by 2. For example, (3, 5), (11, 13), and (17, 19) are twin prime pairs. Twin primes have fascinated mathematicians for centuries, and the Twin Prime Conjecture suggests that there are infinitely many twin primes.

Primality Testing: Determining whether a given number is prime or composite is an important problem in number theory. Various primality testing algorithms have been developed, ranging from simple methods like trial division to more sophisticated algorithms such as the Miller-Rabin test and the AKS primality test.

Prime Number Theorem: The Prime Number Theorem, proven by Jacques Hadamard and Charles Jean de la Vallée-Poussin independently in 1896, provides an asymptotic estimate of the number of primes up to a given limit. It states that as the limit n tends towards infinity, the number of primes less than or equal to n is approximately $n / \ln(n)$, where ln denotes the natural logarithm.

Prime numbers play a crucial role in cryptography, number theory algorithms, and various areas of mathematics. They are the foundation for encryption algorithms like RSA, where security relies on the difficulty of factoring large composite numbers into their prime factors. Prime numbers continue to be a topic of active research, with ongoing efforts to discover large prime numbers and understand their properties and distribution patterns.

3.3 Prime Number Distribution

The distribution of prime numbers refers to how prime numbers are distributed among the positive integers. While the distribution of primes does not follow a predictable pattern, there are some notable characteristics and results regarding their distribution. Here are some key aspects of prime number distribution:

Prime Number Theorem: The Prime Number Theorem, proven by Jacques Hadamard and Charles Jean de la Vallée-Poussin independently in 1896, provides an approximation for the density of prime numbers. It states that as the limit n tends towards infinity, the number of primes less than or equal to n is approximately n / ln(n), where ln denotes the natural logarithm. This theorem gives an estimate of how frequently prime numbers occur as n increases.

Prime Gaps: Prime gaps refer to the differences between consecutive prime numbers. While there are infinitely many prime numbers, the gaps between primes can be arbitrarily large. However, the question of whether there are infinitely many pairs of consecutive primes with a gap of 2, known as twin primes, remains an open problem (the Twin Prime Conjecture). Prime gaps are irregular and exhibit apparent randomness.

Prime Number Races: Certain arithmetic progressions have an unexpectedly high density of prime numbers. These patterns are known as prime number races or prime constellations. Examples include prime quadruplets (prime numbers that are consecutive odd numbers, such as 5, 7, 11, and 13) and prime k-tuples. Prime number races have been studied extensively, and the Green-Tao Theorem (proved in 2004) shows that there are arbitrarily long arithmetic progressions of prime numbers.

Sieve Methods: Sieve methods, such as the Sieve of Eratosthenes, are algorithms used to efficiently identify prime numbers within a given range. These methods rely on eliminating multiples of known primes to isolate the remaining prime numbers. Sieve methods provide insights into the distribution of primes and help identify prime numbers within a specified limit.

Prime Number Bias: While the distribution of prime numbers is generally irregular, certain arithmetic progressions show a bias toward producing primes. This bias is observed in quadratic polynomials such as $n^2 + n + 41$ (Euler's prime-generating polynomial) and $n^2 - n + 41$ (the second Hardy-Littlewood prime-generating polynomial). These polynomials produce prime values for a high number of consecutive integers.

Prime Number Gaps: The prime gaps between consecutive primes tend to increase as the numbers get larger. However, the exact behavior of prime gaps is still not fully understood. The largest known prime gap is the Cramer conjecture gap, which states that there is always a prime gap of size at least $\log^2(n)$ for some constant log.

The distribution of prime numbers remains an active area of research in number theory. While certain patterns and characteristics have been observed, there is still much to discover about the precise nature of prime number distribution. The exploration of prime number distribution is essential for understanding the fundamental properties of prime numbers and their role in various mathematical applications.

3.4 Prime number generation algorithms

Prime number generation algorithms are used to efficiently generate prime numbers within a given range or find large prime numbers. These algorithms leverage various mathematical techniques and properties of prime numbers. Here are some commonly used prime number generation algorithms:

Trial Division: Trial division is the simplest primality testing method. It involves checking each number starting from 2 up to the square root of the candidate number for divisibility. If no divisor is found, the number is prime. While trial division works well for small numbers, it becomes computationally expensive for larger numbers.

Sieve of Eratosthenes: The Sieve of Eratosthenes is an ancient algorithm used to find all prime numbers up to a given limit. It eliminates multiples of known primes to identify the remaining primes. The algorithm starts by marking 2 as prime and crosses out all its multiples. It then moves to the next unmarked number and repeats the process. The Sieve of Eratosthenes is efficient for generating prime numbers within a limited range.

Sieve of Atkin: The Sieve of Atkin is an optimized variation of the Sieve of Eratosthenes. It uses a more refined method to identify prime numbers by considering modulo patterns. The Sieve of Atkin is more efficient than the trial division and the Sieve of Eratosthenes for larger ranges.

Probabilistic Primality Tests: Probabilistic primality tests provide a fast way to determine if a number is likely to be prime. Examples include the Miller-Rabin test and the Fermat primality test. These tests perform multiple iterations of a probabilistic check based on number theory concepts. While they can't guarantee primality with absolute certainty, they are highly accurate and widely used in practice.

Number Theory

Deterministic Primality Tests: Deterministic primality tests, such as the AKS primality test, provide a rigorous and deterministic way to prove the primality of a number. These tests have polynomial-time complexity but are generally slower than probabilistic tests. Deterministic tests are used to rigorously prove the primality of specific numbers.

Prime Number Formulas: Various formulas exist that generate prime numbers under specific conditions. For example, Euler's prime-generating polynomial ($n^2 + n + 41$) produces prime numbers for consecutive integer values of n. Other formulas, such as quadratic sieves and elliptic curve primality proving, use advanced mathematical techniques to generate large prime numbers.

Prime Number Generation with Prime Gaps: Some algorithms focus on generating prime numbers with specific prime gaps, such as twin primes or prime k-tuples. These algorithms explore patterns and properties of prime number races to generate prime numbers with desired properties.

Prime number generation algorithms are essential for applications that require prime numbers, such as cryptography, number theory research, and computer science. These algorithms play a crucial role in finding prime numbers efficiently and exploring the properties and distribution of primes.

4. Modular Arithmetic

Modular arithmetic is a branch of mathematics that deals with the properties and operations of integers within a fixed modulus. It focuses on the remainder of numbers when divided by a specified modulus. Modular arithmetic has applications in various fields, including number theory, cryptography, computer science, and algebraic manipulation. Here are some key concepts and operations in modular arithmetic:

Modulus: The modulus is a fixed positive integer that defines the range of numbers in modular arithmetic. It is denoted by the symbol "m." For example, in modular arithmetic modulo 7, the modulus is 7.

Congruence: Congruence is an equivalence relation between two numbers based on their remainders when divided by the modulus. Two numbers, a and b, are said to be congruent modulo m if they leave the same remainder when divided by m. This is denoted as $a \equiv b \pmod{m}$.

Modular Addition: In modular arithmetic, addition is performed as usual, but the result is reduced modulo m. To add two numbers a and b modulo m, first, calculate a + b, and then take the remainder when divided by m. The result is the congruence class of the sum modulo m.

Modular Subtraction: Similar to modular addition, subtraction is performed as usual, but the result is reduced modulo m. To subtract b from a modulo m, calculate a - b, and then take the remainder when divided by m. The result is the congruence class of the difference modulo m.

Modular Multiplication: Multiplication in modular arithmetic involves multiplying two numbers and then reducing the result modulo m. To multiply a and b modulo m, calculate a * b, and then take the remainder when divided by m. The result is the congruence class of the product modulo m.

Modular Exponentiation: Exponentiation in modular arithmetic is performed by raising a number to a power and reducing the result modulo m. To calculate a^b modulo m, raise a to the power of b and then take the remainder when divided by m.

Modular Inverse: The modular inverse of a number a modulo m is another number x such that (a * x) ≡ 1 (mod m). It represents the multiplicative inverse of an in the modular arithmetic system.

Modular Division: Division is performed by multiplying the dividend by the modular inverse of the divisor. To divide a by b modulo m, calculate (a * x), where x is the modular inverse of b modulo m.

Modular arithmetic provides a way to work with remainders and cyclic patterns in numbers. It is particularly useful in cryptography algorithms like RSA, where modular arithmetic operations ensure the security of encryption and decryption processes. Modular arithmetic also helps in solving equations involving congruences, analyzing number patterns, and optimizing algorithms that work with periodic or repetitive structures.

4.1 Congruence relation

In modular arithmetic, the congruence relation is an equivalence relation that compares two numbers based on their remainders when divided by a specified modulus. Congruence establishes a relationship between numbers that have the same remainder and fall within the same congruence class. The notation used to represent congruence is "≡" (triple bar) or "≡ (mod m)," where "m" is the modulus. Here are the key aspects of the congruence relation:

Number Theory

Definition: Two integers a and b are said to be congruent modulo m if they leave the same remainder when divided by m. In other words, a and b have the same remainder and are part of the same congruence class modulo m. This is denoted as $a \equiv b \pmod{m}$.

Congruence Class: The congruence class, also known as the residue class, is a set of all numbers that are congruent to a given number modulo m. For example, the congruence class of 5 modulo 3 consists of all integers that leave a remainder of 5 when divided by 3, such as 5, 8, 11, -1, -4, and so on.

Equivalence Relation: The congruence relation satisfies the properties of an equivalence relation. It is reflexive, symmetric, and transitive:

- ✓ Reflexive: Every number a is congruent to itself modulo m. ($a \equiv a \pmod{m}$)
- ✓ Symmetric: If a is congruent to b modulo m, then b is congruent to a modulo m. (If $a \equiv b \pmod{m}$, then $b \equiv a \pmod{m}$)
- ✓ Transitive: If a is congruent to b modulo m and b is congruent to c modulo m, then a is congruent to c modulo m. (If $a \equiv b \pmod{m}$ and $b \equiv c \pmod{m}$, then $a \equiv c \pmod{m}$)

Arithmetic Operations: The congruence relation is compatible with arithmetic operations. This means that if $a \equiv b \pmod{m}$ and $c \equiv d \pmod{m}$, then:

- ✓ $a + c \equiv b + d \pmod{m}$ (Congruence is preserved under addition)
- ✓ $a - c \equiv b - d \pmod{m}$ (Congruence is preserved under subtraction)
- ✓ $a * c \equiv b * d \pmod{m}$ (Congruence is preserved under multiplication)

These properties allow us to perform modular arithmetic operations by focusing on congruence classes rather than individual numbers.

Number Theory

The congruence relation plays a crucial role in modular arithmetic, number theory, and cryptography. It helps in analyzing number patterns, solving equations involving congruences, and understanding cyclic behavior in numbers. Congruences are used in various applications, such as encryption algorithms, primality testing, and solving systems of linear congruences.

4.2 Modular addition, subtraction, and multiplication

In modular arithmetic, addition, subtraction, and multiplication are performed similarly to regular arithmetic operations, but with an additional step of reducing the result modulo a specified modulus. This reduction ensures that the final result remains within the congruence class of numbers modulo the modulus. Here's how modular addition, subtraction, and multiplication are carried out:

Modular Addition:

To perform modular addition, add the two numbers as usual and then take the remainder when divided by the modulus. The result is the congruence class of the sum modulo the modulus. The notation used is (a + b) mod m.

Example:

Let's perform modular addition modulo 7.

$5 + 4 \equiv (5 + 4) \mod 7 \equiv 9 \mod 7 \equiv 2 \pmod{7}$

So, $5 + 4 \equiv 2 \pmod{7}$.

Modular Subtraction:

To perform modular subtraction, subtract the second number from the first number as usual and then take the remainder when divided by the modulus. The result is the congruence class of the difference modulo the modulus. The notation used is (a - b) mod m.

Example:

Let's perform modular subtraction modulo 7.

$9 - 4 \equiv (9 - 4) \mod 7 \equiv 5 \mod 7 \equiv 5 \pmod 7$

So, $9 - 4 \equiv 5 \pmod 7$.

Modular Multiplication:

To perform modular multiplication, multiply the two numbers as usual and then take the remainder when divided by the modulus. The result is the congruence class of the product modulo the modulus. The notation used is (a * b) mod m.

Example:

Let's perform modular multiplication modulo 7.

$5 * 4 \equiv (5 * 4) \mod 7 \equiv 20 \mod 7 \equiv 6 \pmod 7$

So, $5 * 4 \equiv 6 \pmod 7$.

Modular arithmetic operations ensure that the results remain within the desired congruence class modulo the specified modulus. This is useful in applications such as cryptography, where calculations are performed within finite fields defined by a modulus. Modular operations also help in solving equations involving congruences and in analyzing cyclic patterns in numbers.

4.3 Modular Exponentiation

Modular exponentiation is the process of raising a number to a power and reducing the result modulo a specified modulus. It is a

Number Theory

fundamental operation in modular arithmetic and finds applications in various areas, including number theory, cryptography, and computer science. Here's how modular exponentiation is performed:

Given three numbers: the base (a), the exponent (b), and the modulus (m), the goal is to calculate a^b modulo m.

1. Start with an initial result of 1: result = 1.

2. Iterate through the binary representation of the exponent, starting from the most significant bit.

3. For each bit, do the following:

   a. Square the current result: result = result * result modulo m. This step is performed regardless of whether the bit is 0 or 1.

   b. If the current bit is 1, multiply the result by the base: result = result * a modulo m.

   After iterating through all the bits of the exponent, the final result is the congruence class of a^b modulo m.

Here's an example to illustrate modular exponentiation:

Let's calculate 3^13 modulo 7.

Binary representation of 13: 1101.

Initialize result = 1.

Starting from the most significant bit:

1. Square the result: result = 1 * 1 modulo 7 = 1.

2. Multiply the result by the base: result = 1 * 3 modulo 7 = 3.

3. Square the result: result = 3 * 3 modulo 7 = 2.

4. Square the result: result = 2 * 2 modulo 7 = 4.

5. Multiply the result by the base: result = 4 * 3 modulo 7 = 5.

6. Square the result: result = 5 * 5 modulo 7 = 4.

7. Multiply the result by the base: result = 4 * 3 modulo 7 = 5.

After iterating through all the bits of the exponent, the result is 5.

Therefore, 3^13 modulo 7 is 5.

Modular exponentiation allows for efficient computation of large exponential expressions modulo a given modulus. It is crucial in various cryptographic algorithms, such as RSA, where the security relies on modular exponentiation operations. Modular exponentiation also plays a significant role in number theory and algorithms involving large numbers and modular arithmetic.

4.4 Modular Inverses

In modular arithmetic, the modular inverse of a number refers to another number that, when multiplied by the original number, yields a congruence of 1 modulo a specified modulus. In other words, given a number a and a modulus m, the modular inverse of a is a number x such that (a * x) ≡ 1 (mod m). The existence of a modular inverse depends on the properties of the number and the modulus. Here's how to find the modular inverse:

Extended Euclidean Algorithm: The extended Euclidean algorithm is commonly used to find the modular inverse. It determines the greatest common divisor (GCD) of two numbers and provides coefficients that satisfy a linear combination equation. In the context of modular

Number Theory

inverses, the extended Euclidean algorithm helps find the coefficients that produce a GCD of 1.

Modular Inverse Calculation: To find the modular inverse of a number a modulo m using the extended Euclidean algorithm, follow these steps:

   a. Apply the extended Euclidean algorithm to a and m. This will give you the GCD of a and m and the coefficients x and y such that ax + my = GCD (a, m) = 1.

   b. Take the coefficient x obtained from the extended Euclidean algorithm. It is the modular inverse of a modulo m.

   c. If x is negative, add m to it to get a positive modular inverse.

   Note: If the GCD of a and m is not 1, the modular inverse does not exist.

Here's an example to illustrate finding the modular inverse:

Let's find the modular inverse of 5 modulo 13.

Using the extended Euclidean algorithm:

$13 = 2 * 5 + 3$

$5 = 1 * 3 + 2$

$3 = 1 * 2 + 1$

$2 = 2 * 1 + 0$

Working backward, we have:

$1 = 3 - 1 * 2$

$= 3 - 1 * (5 - 1 * 3)$

$= 2 * 3 - 1 * 5$

$= 2 * (13 - 2 * 5) - 1 * 5$

$= 2 * 13 - 5 * 5$

The coefficient of 5 is -5, which is the modular inverse.

Since -5 is negative, we add 13 to it:

The modular inverse of 5 modulo 13 = -5 + 13 = 8.

Therefore, the modular inverse of 5 modulo 13 is 8.

Modular inverses are essential in modular arithmetic, especially in applications such as modular division and cryptography. They allow for the computation of multiplicative inverses within a modular system. It's important to note that not all numbers have a modular inverse. If the GCD of the number and the modulus is not 1, the modular inverse does not exist.

Number Theory

5. Diophantine Equations

Diophantine equations are algebraic equations that involve integer solutions. They are named after the ancient Greek mathematician Diophantus of Alexandria, who studied these types of equations extensively. Diophantine equations are distinct from other types of equations because they require integer solutions rather than real or rational solutions. Here's an overview of Diophantine equations:

Form: A Diophantine equation is typically written in the form:

$f(x_1, x_2, ..., x_n) = 0$,

where f is a polynomial with integer coefficients, and $x_1, x_2, ..., x_n$ are integer variables.

Integer Solutions: The goal of solving a Diophantine equation is to find all possible integer solutions $(x_1, x_2, ..., x_n)$ that satisfy the equation. These solutions must be integers rather than real or rational numbers.

Types of Diophantine Equations: Diophantine equations can have various forms and properties. Some common types include:

- ✓ Linear Diophantine Equations: Equations of the form $ax + by = c$, where a, b, and c are integers, and x and y are integer variables.
- ✓ Homogeneous Diophantine Equations: Equations of the form $ax + by = 0$, where a and b are integers, and x and y are integer variables.
- ✓ Pell's Equations: Equations of the form $x^2 - dy^2 = 1$, where d is a positive integer.
- ✓ Fermat's Last Theorem: The equation $x^n + y^n = z^n$, where n is an integer greater than 2. This equation remained unsolved for centuries until Andrew Wiles proved it in 1994.

Methods of Solution: Solving Diophantine equations can be challenging, and in many cases, there is no general algorithm to find all solutions. Different methods and techniques are used based on the specific form of the equation. Some common approaches include:

- ✓ Brute force: Checking all possible integer values within a given range to find solutions.
- ✓ Linear Diophantine Equation Solvers: Using techniques such as the Extended Euclidean Algorithm to find specific solutions or all solutions in a given range.
- ✓ Number Theory Techniques: Utilizing concepts from number theory, such as modular arithmetic, prime factorization, and congruences, to analyze and solve specific Diophantine equations.

Applications: Diophantine equations have applications in various fields, including number theory, cryptography, computer science, and optimization problems. They are used to model and solve problems that require integer solutions.

Solving Diophantine equations often involves a combination of mathematical analysis, number theory techniques, and creative problem-solving. While some equations have known general solutions, many Diophantine equations remain unsolved or require specialized methods for specific cases. The study of Diophantine equations continues to be an active area of research, with ongoing efforts to develop new algorithms and techniques for solving these equations.

5.1 Linear Diophantine Equations

Linear Diophantine equations are a specific type of Diophantine equation in which the variables are linearly related. They can be written in the form $ax + by = c$, where a, b, and c are integers, and x and y are integer variables. The goal is to find integer solutions for x and y that satisfy the equation. Here's an overview of solving linear Diophantine equations:

GCD Condition: The existence of integer solutions for a linear Diophantine equation $ax + by = c$ depends on the greatest common divisor (GCD) of a and b. If the GCD of a and b divides c (i.e., $GCD(a, b) \mid c$), then integer solutions exist. If not, the equation has no integer solutions.

Extended Euclidean Algorithm: The Extended Euclidean Algorithm is commonly used to find the GCD of two integers (a and b) and express it as a linear combination of a and b. By applying the Extended Euclidean Algorithm to a and b, you can find integer coefficients (x_0, y_0) such that $ax_0 + by_0 = GCD(a, b)$.

Finding Particular Solution: Once you have the GCD and the coefficients (x_0, y_0), you can find a particular solution to the linear Diophantine equation. Multiply both sides of the equation by $c/GCD(a, b)$ to obtain:

$$a(x_0 * (c/GCD(a, b))) + b(y_0 * (c/GCD(a, b))) = c.$$

The values $x = x_0 * (c/GCD(a, b))$ and $y = y_0 * (c/GCD(a, b))$ form a particular solution to the linear Diophantine equation.

General Solution: The general solution to a linear Diophantine equation is obtained by adding integer multiples of $b/GCD(a, b)$ to x and subtracting integer multiples of $a/GCD(a, b)$ from y. The general solution can be written as:

$$x = x_0 * (c/GCD(a, b)) + (b/GCD(a, b)) * t,$$

$$y = y_0 * (c/GCD(a, b)) - (a/GCD(a, b)) * t,$$

where t is an integer representing the parameter that generates all possible integer solutions.

Number Theory

Range of Solutions: Depending on the problem's constraints, you might need to find solutions within a specific range. You can set bounds on the parameter t to limit the values of x and y within a desired range.

By following these steps, you can find integer solutions to linear Diophantine equations. Linear Diophantine equations have applications in various areas, including number theory, cryptography, and optimization problems. They can be solved using number theory techniques, such as the Extended Euclidean Algorithm, modular arithmetic, and concepts from linear algebra.

5.2 Pythagorean Triples

Pythagorean triples are sets of three positive integers (a, b, c) that satisfy the Pythagorean theorem, a fundamental result in geometry. The Pythagorean theorem states that in a right-angled triangle, the square of the length of the hypotenuse (the side opposite the right angle) is equal to the sum of the squares of the other two sides. Here's an overview of Pythagorean triples:

Definition: A Pythagorean triple consists of three positive integers (a, b, c) that satisfy the equation $a^2 + b^2 = c^2$.

Primitive Pythagorean Triples: A Pythagorean triple (a, b, c) is called primitive if a, b, and c have no common factors (i.e., they are coprime). In a primitive Pythagorean triple, a and b are both odd or one of them is divisible by 3.

Generating Pythagorean Triples: There are various methods to generate Pythagorean triples. Some common approaches include:

a. Euclid's Formula: Euclid's formula generates all primitive Pythagorean triples. It states that for any two positive integers m and

Number Theory

n (m > n), the following values of a, b, and c will form a primitive Pythagorean triple:

$a = m^2 - n^2,$

$b = 2mn,$

$c = m^2 + n^2.$

The values of m and n determine different Pythagorean triples.

b. Pythagorean Triple Identities: There are several identities that can be used to generate new Pythagorean triples from existing ones. For example, multiplying a known triple by a constant, interchanging the values of a and b, or scaling a triple by a common factor can generate new triples.

c. Generating Triples from Primitive Triples: Non-primitive Pythagorean triples can be obtained by multiplying the values of a, b, and c in a primitive triple by a common factor.

Examples of Pythagorean Triples

- (3, 4, 5) is the most well-known Pythagorean triple.

- (5, 12, 13), (8, 15, 17), (7, 24, 25), and (20, 21, 29) are other examples of Pythagorean triples.

Applications: Pythagorean triples have applications in geometry, number theory, and practical problems involving right-angled triangles. They are used in the construction of right-angled triangles, determining side lengths, and solving problems involving integer side lengths of triangles.

Pythagorean triples are a fascinating aspect of number theory, and their properties continue to be studied and explored. They provide valuable insights into the relationships between integers and the geometry of right-angled triangles.

5.3 Fermat's Last Theorem

Fermat's Last Theorem is one of the most famous and intriguing theorems in the history of mathematics. It states that there are no three positive integers a, b, and c that satisfy the equation $a^n + b^n = c^n$ for any integer value of n greater than 2. In other words, there are no non-zero integer solutions to the equation when n is an integer greater than 2.

Fermat's Last Theorem was proposed by the French mathematician Pierre de Fermat in the 17th century. Fermat himself claimed to have proof for the theorem but famously wrote in the margin of his notebook that the margin was too small to contain it. For centuries, mathematicians attempted to prove or disprove Fermat's Last Theorem, but it remained an unsolved problem.

The quest to prove Fermat's Last Theorem continued until 1994 when the British mathematician Andrew Wiles presented proof for the theorem. Wiles' proof involved a deep and intricate connection between modular forms, elliptic curves, and Galois representations. The proof was groundbreaking and involved advanced mathematical concepts and techniques.

Wiles' proof of Fermat's Last Theorem marked a significant milestone in the history of mathematics. It demonstrated the power of mathematical reasoning and highlighted the complexity and depth of number theory. The proof was reviewed and accepted by the mathematical community, and it resolved one of the most enduring problems in mathematics.

Fermat's Last Theorem has had a profound impact on the field of number theory and related areas of mathematics. It has spurred further research and exploration into related topics, such as elliptic curves, modular forms, and Galois representations. The techniques developed to prove Fermat's Last Theorem have found applications in various branches of mathematics and have influenced the development of new mathematical ideas and theories.

Fermat's Last Theorem stands as a testament to the power of mathematical inquiry and the pursuit of knowledge. It reminds us of the richness and depth of mathematics and the continuous quest to explore and understand the fundamental truths of the mathematical universe.

6. Euler's Totient Function

Euler's totient function, also known as Euler's phi function or simply the totient function, is a number theory function that counts the positive integers up to a given number that are relatively prime to it. The totient function is denoted by the symbol $\varphi(n)$, where n is a positive integer. Euler's totient function has several important properties and applications in number theory and cryptography. Here's an overview of Euler's totient function:

Definition: The totient function $\varphi(n)$ counts the positive integers less than or equal to n that are coprime (relatively prime) to n. In other words, $\varphi(n)$ gives the count of numbers between 1 and n (inclusive) that have no common factors with n, except for 1.

Formula: Euler's totient function can be calculated using a formula based on the prime factorization of n. If n can be expressed as a product of distinct prime factors $p_1^{k_1} * p_2^{k_2} * ... * p_n^{k_n}$, then the totient function is given by:

$$\varphi(n) = n * (1 - 1/p_1) * (1 - 1/p_2) * ... * (1 - 1/p_n).$$

Properties:

- If p is a prime number, then $\varphi(p) = p - 1$, as all numbers from 1 to p - 1 are coprime to p.

- If a and b are coprime, then $\varphi(ab) = \varphi(a) * \varphi(b)$. This property is known as the multiplicative property of the totient function.

- If n is a prime power, i.e., $n = p^k$ for some prime p and positive integer k, then $\varphi(n) = p^k - p^{(k-1)}$.

Applications:

- ✓ Primality Testing: Euler's totient function is used in primality testing algorithms like the Miller-Rabin test and the Fermat primality test.
- ✓ Cryptography: Euler's totient function plays a crucial role in public key encryption algorithms such as RSA. It is used to calculate the public and private keys and to ensure the security of the encryption scheme.
- ✓ Permutation Groups: Euler's totient function is used in counting the number of elements in certain permutation groups.

Euler's totient function provides valuable insights into the properties of numbers and their relationships with coprime integers. It has applications in various areas of number theory, cryptography, and combinatorics. The study of the totient function continues to be an active area of research, with ongoing efforts to explore its properties, generalize its formulas, and discover new applications.

6.1 Definition and Properties

Euler's totient function, denoted as $\varphi(n)$, is a number theory function that counts the positive integers up to a given number n that are coprime (relatively prime) to n. In other words, it calculates the count of numbers between 1 and n (inclusive) that have no common factors with n except for 1.

Definition: Euler's totient function $\varphi(n)$ is defined as the count of positive integers less than or equal to n that are coprime to n. It measures the number of positive integers in the range 1 to n (inclusive) that share no common factors with n except for 1.

Here's a closer look at the properties of Euler's totient function:

Coprime (Relatively Prime): Two numbers a and b are said to be coprime or relatively prime if their greatest common divisor (GCD) is 1. In other words, a and b have no common factors except for 1. For example, if n = 10, the positive integers coprime to 10 are 1, 3, 7, and 9, as these numbers do not share any factors with 10 except for 1.

Prime Numbers: If n is a prime number p, then $\varphi(p) = p - 1$, as all positive integers less than p are coprime to p.

Multiplicative Property: Euler's totient function has a useful multiplicative property. If a and b are coprime positive integers, then $\varphi(ab) = \varphi(a) * \varphi(b)$. In other words, the totient function of a product is equal to the product of the totient functions of the individual numbers. This property allows the totient function to be easily calculated for composite numbers based on their prime factorization.

Prime Power: If n is a prime power, i.e., $n = p^k$ for some prime number p and positive integer k, then $\varphi(n) = p^k - p^{k-1}$. This property holds because a number is coprime to p^k if and only if it is not divisible by p.

Summation Property: The sum of $\varphi(d)$ for all positive divisors d of n is equal to n. In other words, if d divides n, then the values of $\varphi(d)$ for all divisors d of n, when summed together, will give the value of n.

Relationship with Euler's Totient Theorem: Euler's Totient Theorem states that for any positive integer n coprime to a positive integer a, $a^{\varphi(n)} \equiv 1 \pmod{n}$, where $\equiv$ denotes congruence modulo n. This theorem relates the totient function to the concept of modular exponentiation.

Euler's totient function has numerous applications in number theory, cryptography, and combinatorics. Its properties help in understanding the relationships between numbers and their coprimality, and they provide valuable insights into the properties of

Number Theory

integers and their factors. The totient function is widely used in algorithms and mathematical calculations, making it a fundamental concept in number theory.

6.2 Euler's Theorem

I apologize for the confusion in the previous response. Euler's theorem is not directly related to the totient function. It is actually a special case of Euler's totient theorem, which is a significant result in number theory. Let's discuss Euler's theorem and its connection to Euler's totient theorem:

Euler's Theorem:

Euler's theorem states that for any positive integers a and n that are coprime (i.e., gcd(a, n) = 1), $a^{\varphi(n)} \equiv 1 \pmod{n}$. Here, $\varphi(n)$ represents Euler's totient function, which calculates the count of positive integers coprime to n. The notation $\equiv$ denotes congruence modulo n.

Key Points:

Coprime Integers: For Euler's theorem to hold, a and n must be coprime, meaning they have no common factors except for 1. This ensures the application of Euler's totient theorem.

Modular Congruence: The congruence $a^{\varphi(n)} \equiv 1 \pmod{n}$ means that $a^{\varphi(n)}$ leaves a remainder of 1 when divided by n. In other words, $a^{\varphi(n)}$ is congruent to 1 modulo n.

Modular Exponentiation: Euler's theorem demonstrates a special property of modular exponentiation when the base and modulus are coprime. The result is always congruent to 1 modulo the modulus.

Euler's Totient Theorem:

Euler's totient theorem is a more general result, from which Euler's theorem is derived. Euler's totient theorem states that if a and n are coprime positive integers, then $a^{\varphi(n)} \equiv 1 \pmod{n}$. The key difference is that Euler's totient theorem holds for any coprime pair (a, n), while Euler's theorem specifically deals with the case when a and n are coprime and a is raised to the power of $\varphi(n)$.

The connection between Euler's Theorem and Euler's Totient Theorem:

Euler's theorem is a special case of Euler's totient theorem. When a and n are coprime, the totient function $\varphi(n)$ evaluates to the count of positive integers coprime to n. Therefore, $a^{\varphi(n)} \equiv 1 \pmod{n}$ becomes $a^{count} \equiv 1 \pmod{n}$, which is Euler's theorem.

Euler's theorem and Euler's totient theorem have profound applications in number theory and cryptography, particularly in the field of modular arithmetic. They are fundamental results that provide insights into the properties of coprime integers and the behavior of exponentiation in modular arithmetic. These theorems form the basis for various cryptographic algorithms and are widely used in number theoretic computations.

6.3 Applications in Cryptography

Euler's totient theorem and Euler's theorem have significant applications in the field of cryptography. They play a crucial role in several cryptographic algorithms and protocols. Here are some notable applications:

RSA Encryption: The RSA (Rivest-Shamir-Adleman) encryption algorithm is one of the most widely used public-key encryption schemes. Euler's totient theorem is at the heart of RSA. The theorem ensures the security of RSA by utilizing the difficulty of factoring large numbers. The totient function φ(n) is used to calculate the public and private keys in RSA, where n is a product of two large prime numbers.

Primality Testing: Euler's totient theorem is employed in primality testing algorithms. For instance, the Fermat primality test and the Miller-Rabin primality test use modular exponentiation, which relies on Euler's theorem. These tests determine whether a given number is prime with high probability.

Diffie-Hellman Key Exchange: The Diffie-Hellman key exchange protocol is a widely used method for secure key exchange between two parties over an insecure communication channel. Euler's theorem is utilized in this protocol to perform modular exponentiation operations and establish a shared secret key between the parties.

ElGamal Encryption: The ElGamal encryption scheme is another public-key encryption algorithm that relies on the difficulty of computing discrete logarithms. It employs Euler's totient theorem to perform modular exponentiation and generate the ciphertext.

Digital Signatures: Digital signature algorithms, such as the Digital Signature Algorithm (DSA) and Elliptic Curve Digital Signature Algorithm (ECDSA), utilize modular exponentiation and the properties of Euler's totient theorem to sign and verify digital signatures. The security of these algorithms is closely tied to the difficulty of solving discrete logarithm problems.

Cryptographic Protocols: Euler's totient theorem and its applications in modular arithmetic are used in various cryptographic protocols and systems. These include key exchange protocols, cryptographic hash functions, secure multiparty computation, and more.

Euler's totient theorem and Euler's theorem provide the foundation for many cryptographic techniques, ensuring the security and integrity of data transmission, authentication, and privacy in modern digital communication. These theorems leverage the properties of modular arithmetic and the difficulty of certain mathematical problems to enable secure cryptographic operations.

7. Quadratic Residues

Quadratic residues are an important concept in number theory and modular arithmetic. They are closely related to quadratic congruences and have various applications in fields such as cryptography and number theory. Let's explore quadratic residues and their properties:

Definition: Given an integer modulus m and an integer a, a quadratic residue modulo m is an integer x such that $x^2 \equiv a \pmod{m}$. In other words, a quadratic residue is a value of x that, when squared and reduced modulo m, yields the value of a.

Quadratic Non-residue: If there is no integer x such that $x^2 \equiv a \pmod{m}$, then a is called a quadratic non-residue modulo m. In this case, there are no solutions for the congruence $x^2 \equiv a \pmod{m}$.

Quadratic Residue Classes: The set of all quadratic residues modulo m forms a subgroup of the integers modulo m. This subgroup is called the quadratic residue class modulo m. The number of quadratic residues modulo m depends on the properties of the modulus m.

Properties of Quadratic Residues:

- ✓ If x is a quadratic residue modulo m, then (-x) is also a quadratic residue modulo m.
- ✓ The product of two quadratic residues modulo m is always a quadratic residue modulo m.
- ✓ The product of a quadratic residue and a quadratic non-residue modulo m is a quadratic non-residue modulo m.

Quadratic Reciprocity: Quadratic reciprocity is a fundamental result in number theory that provides a relationship between the quadratic residues modulo two different prime moduli. The quadratic reciprocity theorem, formulated by Carl Friedrich Gauss, establishes a criterion for determining when a prime p is a quadratic residue modulo another prime q, and vice versa.

Applications: Quadratic residues have applications in various areas, including cryptography, pseudorandom number generation, and primality testing. They play a critical role in cryptographic schemes like the Quadratic Residuosity Assumption, which is used in constructing secure cryptographic protocols.

Studying quadratic residues helps in understanding the behavior of quadratic congruences and the structure of the quadratic residue class modulo a given modulus. The properties and relationships of quadratic residues are fundamental to many advanced topics in number theory, algebraic structures, and cryptography.

7.1 Quadratic residues and non-residues

Quadratic residues and non-residues are distinct classes of integers with respect to a given modulus. Understanding their properties and relationships is essential in number theory and modular arithmetic. Let's delve deeper into quadratic residues and non-residues:

Quadratic Residues: Given a modulus m, a quadratic residue modulo m is an integer a such that there exists an integer x satisfying $x^2 \equiv a$ (mod m). In other words, a quadratic residue is a value that can be obtained by squaring an integer and reducing it modulo m. Quadratic residues form a subset of the integers modulo m and have certain properties associated with them.

Quadratic Non-residues: Conversely, a quadratic non-residue modulo m is an integer a for which there is no integer x satisfying $x^2 \equiv a$ (mod m). Quadratic non-residues cannot be obtained by squaring any integer and reducing it modulo m. Similar to quadratic residues, quadratic non-residues also form a subset of the integers modulo m.

Number Theory

Cardinality: The number of quadratic residues modulo m depends on the properties of the modulus m. If m is an odd prime, there are (m-1)/2 quadratic residues and (m-1)/2 quadratic non-residues modulo m. If m is composite, the number of quadratic residues can be less than (m-1)/2.

Properties:

- ✓ If a is a quadratic residue modulo m, then (-a) is also a quadratic residue modulo m.
- ✓ The product of two quadratic residues modulo m is always a quadratic residue modulo m.
- ✓ The product of a quadratic residue and a quadratic non-residue modulo m is a quadratic non-residue modulo m.

Quadratic Reciprocity: Quadratic reciprocity is a fundamental result in number theory that establishes a relationship between quadratic residues modulo two different primes. The quadratic reciprocity theorem, formulated by Carl Friedrich Gauss, provides conditions for determining when a prime p is a quadratic residue modulo another prime q, and vice versa.

Applications: Quadratic residues and non-residues have various applications in cryptography, number theory, and pseudorandom number generation. They are used in cryptographic protocols, primality testing algorithms, and generating pseudorandom numbers with desired properties.

The study of quadratic residues and non-residues allows for a deeper understanding of the behavior of quadratic congruences and their relation to modular arithmetic. These concepts are essential in various fields of mathematics, particularly in number theory and cryptography, where they play a crucial role in developing secure algorithms and protocols.

7.2 Quadratic Reciprocity Theorem

The quadratic reciprocity theorem is a significant result in number theory that establishes a relationship between quadratic residues modulo two different prime numbers. It provides a criterion for determining whether a prime number is a quadratic residue modulo another prime. The theorem was formulated by Carl Friedrich Gauss and is widely regarded as one of the most important results in number theory. Here's an overview of the quadratic reciprocity theorem:

Statement of the Quadratic Reciprocity Theorem:

The quadratic reciprocity theorem states that for any two distinct odd prime numbers p and q, the congruence $x^2 \equiv p \pmod{q}$ has a solution if and only if the congruence $x^2 \equiv q \pmod{p}$ has a solution.

In other words, if p and q are distinct odd primes, then p is a quadratic residue modulo q if and only if q is a quadratic residue modulo p.

Symbolically, the quadratic reciprocity theorem can be written as:

Legendre symbol: $(p/q) * (q/p) = (-1)^{((p-1)/2 * (q-1)/2)}$

Jacobi symbol (extended version): $(p/q) = (-1)^{((p-1)/2 * (q-1)/2)}$

Key Points:

Quadratic Residues: The theorem relates the existence of solutions to the congruences $x^2 \equiv p \pmod{q}$ and $x^2 \equiv q \pmod{p}$ to whether p and q are quadratic residues modulo each other.

Reciprocity: The theorem establishes a reciprocal relationship between the quadratic residues of two distinct primes.

Number Theory

Conditions: The theorem applies to distinct odd prime numbers. For the special cases of p = 2 or q = 2, other rules govern the quadratic residues modulo 2.

Derivations: Several proofs of the quadratic reciprocity theorem have been developed over the years, including the original proof by Gauss and subsequent proofs by other mathematicians using different techniques.

Extensions: The quadratic reciprocity theorem has been extended to consider composite numbers, resulting in the development of the Jacobi symbol, which is a generalization of the Legendre symbol.

The quadratic reciprocity theorem has profound implications in number theory and has numerous applications in cryptography, primality testing, and modular arithmetic. It provides insights into the distribution and properties of quadratic residues modulo prime numbers and has paved the way for the development of various number-theoretic algorithms and cryptographic protocols.

7.3 Legendre and Jacobi Symbols

Legendre and Jacobi symbols are mathematical notations used to determine the quadratic character of an integer with respect to a modulus. They are especially useful in studying quadratic residues and non-residues and in applications such as primality testing and cryptographic algorithms. Let's explore Legendre and Jacobi symbols:

Legendre Symbol: The Legendre symbol is defined for an integer a and an odd prime modulus p. It is denoted as (a/p) and is defined as follows:

- (a/p) = 0 if a is divisible by p.

- $(a/p) = 1$ if a is a quadratic residue modulo p (i.e., there exists an integer x such that $x^2 \equiv a \pmod{p}$).

- $(a/p) = -1$ if a is a quadratic non-residue modulo p (i.e., there is no integer x satisfying $x^2 \equiv a \pmod{p}$).

Key Properties of the Legendre Symbol:

- If $a \equiv b \pmod{p}$, then $(a/p) = (b/p)$.

- $(ab/p) = (a/p)(b/p)$ for any integers a and b.

- $(a/p) \equiv a^{((p-1)/2)} \pmod{p}$, known as Euler's criterion.

Jacobi Symbol: The Jacobi symbol extends the concept of the Legendre symbol to include composite moduli. It is denoted as (a/n), where a is an integer and n is a positive odd integer. The Jacobi symbol is defined as follows:

- $(a/n) = 0$ if a is divisible by n.

- $(a/n) = 1$ if a is a quadratic residue modulo n.

- $(a/n) = -1$ if a is a quadratic non-residue modulo n.

Key Properties of the Jacobi Symbol:

- $(a/n) = (a/p_1)(a/p_2)...(a/p_m)$ for a composite modulus $n = p_1^{e_1} * p_2^{e_2} * ... * p_m^{e_m}$, where $p_1, p_2, ..., p_m$ are distinct odd primes.

- $(ab/n) = (a/n)(b/n)$ for any integers a and b.

- The Jacobi symbol is related to the Legendre symbol as follows: $(a/p) = (a/p_1)(a/p_2)...(a/p_m)$ for a prime modulus p.

The Legendre and Jacobi symbols are valuable tools in number theory and modular arithmetic. They help determine the quadratic character of integers and are used in primality testing algorithms like the Solovay-Strassen test and the Miller-Rabin test. Additionally, they are employed in cryptographic protocols involving modular exponentiation and key generation. The properties and computational properties of these symbols make them essential in number theory computations and related applications.

8. Continued Fractions

Continued fractions are a method of representing real numbers as an infinite sequence of nested fractions. They provide an alternative way to express numbers and offer a compact and unique representation of many irrational numbers. Continued fractions have numerous applications in number theory, approximation theory, and algorithm design.

8.1 Definition, Properties, and Applications

Definition: A continued fraction is an expression of the form:

a0 + 1 / (a1 + 1 / (a2 + 1 / (a3 + 1 / (...))))

where a0 is an integer part (or whole number) and the subsequent terms (a1, a2, a3, ...) are positive integers called partial quotients.

Certainly! Let's dive into the *properties* of continued fractions:

Convergents: The convergents of a continued fraction are the partial sums obtained by truncating the infinite sequence at various positions. The nth convergent is denoted as [a0; a1, a2, ..., an]. The convergents provide increasingly accurate rational approximations to the original number represented by the continued fraction.

Rational Numbers: Every rational number can be expressed as a finite continued fraction. For example, the rational number 3/5 can be represented as [0; 1, 2], while 7/4 can be expressed as [1; 1, 1].

Irrational Numbers: Most irrational numbers have infinite continued fraction representations. The continued fraction expansion of an irrational number exhibits a repeating pattern or follows a specific sequence of partial quotients.

Simple Continued Fractions: A simple continued fraction is a continued fraction where all the partial quotients (except possibly the first one) are equal to 1. For example, the square root of 2 can be represented as [1; 2, 2, 2, ...]. Simple continued fractions have special properties and relationships with quadratic irrational numbers.

Unique Representation: Every real number, except for certain special cases like rational numbers, has a unique continued fraction representation. This uniqueness makes continued fractions useful in various applications, such as rational approximation and solving Diophantine equations.

Best Rational Approximations: The convergents of a continued fraction provide the best rational approximations to the original number. The accuracy of the approximations improves as more terms are considered.

Continued Fraction Algorithm: There are algorithms available to compute the continued fraction representation of a given number and to find rational approximations using its convergents. These algorithms enable efficient calculations and manipulations of continued fractions.

Continued fractions possess interesting properties and relationships with various mathematical concepts, including number theory, approximation theory, and quadratic irrational numbers. They have practical applications in cryptography, number theory, and algorithms involving rational approximation. The unique representation and accurate rational approximations provided by continued fractions make them valuable tools in mathematical analysis and problem-solving.

Number Theory

Applications:

- ✓ Rational Approximation: Continued fractions provide a systematic method for finding the best rational approximations to real numbers.
- ✓ Diophantine Equations: Continued fractions can be used to find integer solutions to certain types of Diophantine equations.
- ✓ Number Theory: Continued fractions have connections to various topics in number theory, such as Pell's equation and quadratic forms.
- ✓ Cryptography: Continued fractions have applications in certain cryptographic algorithms, particularly those involving number factorization.

Continued fractions offer an elegant and powerful way to represent real numbers and approximate them with rational numbers. They have deep connections to number theory and provide valuable insights into the properties of real numbers and their relationships. Continued fractions have been extensively studied and applied in various mathematical disciplines, making them a fascinating and important topic in mathematics.

8.2 Convergents and Approximations

Convergents are the partial sums obtained by truncating a continued fraction at various positions. They provide increasingly accurate rational approximations to the original number represented by the continued fraction. Here's a closer look at convergents and their role in approximating real numbers:

Convergents: In a continued fraction [a0; a1, a2, ..., an], the nth convergent is denoted as [a0; a1, a2, ..., an]. It represents the sum of the terms up to the nth position. Convergents are expressed as fractions, with a numerator and denominator.

Number Theory

Rational Approximations: Convergents of continued fractions provide rational approximations to the original number represented by the continued fraction. The accuracy of the approximation improves as more terms are considered.

Calculation of Convergents: The convergents can be calculated iteratively based on the continued fraction expression. The process starts with the integer part a0 and continues by successively adding the partial quotient terms and inverting the resulting sum.

Best Rational Approximations: The convergents of a continued fraction are the best rational approximations to the original number within the set of fractions with a limited denominator size. As the terms in the continued fraction increase, the convergents become better approximations.

Error Bounds: The difference between the original number and its nth convergent is called the error term. The error term becomes smaller as n increases, indicating the improved accuracy of the approximation. The error term can be bounded to estimate the precision of the approximation.

Continued Fraction Algorithm: An algorithm called the continued fraction algorithm is used to calculate convergents and generate rational approximations from a given continued fraction expression. This algorithm allows for efficient computations and enables the use of continued fractions in practical applications.

Best Rational Approximation Property: The convergent with the largest denominator is often considered the best approximation among the convergents. It is the closest rational approximation to the original number within the set of fractions with the same or smaller denominator.

Convergents play a crucial role in providing rational approximations to real numbers represented by continued fractions. They allow for efficient and accurate approximations, making continued fractions a powerful tool for numerical calculations and rational approximation methods. The convergents provide a sequence of increasingly precise rational approximations that capture the value of the original number.

8.3 Pell's Equation

Pell's equation is a type of Diophantine equation that takes the form $x^2 - Dy^2 = 1$, where D is a positive non-square integer. It is named after the English mathematician John Pell, who made significant contributions to its study. Pell's equation has been a subject of investigation for centuries and has various interesting properties. Let's delve into Pell's equation:

General Form: Pell's equation is represented as $x^2 - Dy^2 = 1$, where x and y are positive integers and D is a positive non-square integer. The equation seeks integer solutions that satisfy the equation.

Pell Numbers: The solutions of Pell's equation generate a sequence of pairs (x, y) that are known as Pell numbers. The Pell numbers have a recursive relationship and can be obtained by repeatedly applying the formula:

$x_{n+1} = x_1 * x_n + D * y_1 * y_n$

$y_{n+1} = x_1 * y_n + y_1 * x_n$

where (x_1, y_1) is the fundamental solution of Pell's equation (i.e., the smallest positive integer solution).

Fundamental Solution: Pell's equation always has at least one solution, known as the fundamental solution. The fundamental solution provides the smallest positive integer values (x, y) that satisfy the equation. From the fundamental solution, the entire sequence of solutions (Pell numbers) can be generated.

Infinite Solutions: Pell's equation typically has infinitely many solutions, which form an infinite sequence of pairs (x, y) that satisfy the equation. These solutions can be obtained by iteratively applying the recursive formula mentioned above.

Pell-Lucas Numbers: A variant of Pell's equation, known as the Pell-Lucas equation, takes the form $x^2 - Dy^2 = -1$. This equation generates another sequence of pairs (x, y) called Pell-Lucas numbers.

Continued Fraction Connection: Pell's equation is closely related to continued fractions. The convergents of the continued fraction expansion of sqrt(D) provide increasingly accurate rational approximations to the solutions of Pell's equation.

Applications: Pell's equation has applications in number theory, algebraic number theory, and cryptography. It is used in various algorithms, such as factorization methods, primality testing, and certain cryptographic schemes.

Pell's equation is a fascinating topic in number theory with rich mathematical properties. It continues to be actively researched, and various methods have been developed to solve and analyze its solutions. The equation's connection to continued fractions and its applications in diverse mathematical fields make it an intriguing subject for exploration and study.

9. Cryptography

Cryptography is the science and practice of securing communication and data from unauthorized access or modification. It involves various techniques and algorithms for encoding information, ensuring confidentiality, integrity, authentication, and non-repudiation. Cryptography plays a crucial role in modern information security and has applications in various fields, including online transactions, data protection, secure communication, and more. Let's explore cryptography in more detail:

Encryption and Decryption: Encryption is the process of converting plaintext (original message) into ciphertext (encrypted message) using an encryption algorithm and a secret key. Decryption is the reverse process of converting ciphertext back into plaintext using a decryption algorithm and the same secret key. Encryption ensures that even if an unauthorized person gains access to the ciphertext, they cannot understand its content without the decryption key.

Symmetric Encryption: Symmetric encryption uses a single key for both encryption and decryption. The same secret key is used by both the sender and the recipient to encrypt and decrypt the message. Popular symmetric encryption algorithms include AES (Advanced Encryption Standard), DES (Data Encryption Standard), and 3DES (Triple Data Encryption Standard).

Asymmetric Encryption: Asymmetric encryption, also known as public-key encryption, uses a pair of mathematically related keys: a public key for encryption and a private key for decryption. The public key is widely distributed, while the private key is kept secret by the owner. Asymmetric encryption allows secure communication between parties without needing to exchange a shared secret key. Examples of asymmetric encryption algorithms include RSA (Rivest-Shamir-Adleman) and Elliptic Curve Cryptography (ECC).

Number Theory

Hash Functions: Hash functions are algorithms that transform input data into a fixed-size output called a hash or digest. Hash functions have various applications in cryptography, such as data integrity verification, password storage, and digital signatures. Commonly used hash functions include SHA-256 (Secure Hash Algorithm 256-bit) and MD5 (Message Digest 5).

Digital Signatures: Digital signatures provide a means to authenticate the integrity and origin of digital messages or documents. They use asymmetric encryption techniques to create a unique signature that can be verified using the signer's public key. Digital signatures ensure non-repudiation, meaning the signer cannot deny their association with the signed data. Algorithms like RSA and DSA (Digital Signature Algorithm) are commonly used for digital signatures.

Key Exchange: Key exchange protocols enable two parties to securely establish a shared secret key over an insecure communication channel. The Diffie-Hellman key exchange is a widely used protocol that allows secure key agreement between two parties without transmitting the shared secret key.

Cryptographic Hashing: Cryptographic hashing is the process of transforming input data into a fixed-size hash value using a cryptographic hash function. It is used for password storage, data integrity verification, and message authentication codes (MACs).

Cryptanalysis: Cryptanalysis is the study of breaking cryptographic algorithms and systems. Cryptanalysts analyze cryptographic algorithms and attempt to find weaknesses or vulnerabilities that can be exploited to compromise the security of encrypted data. Cryptanalysis techniques include brute-force attacks, statistical analysis, and mathematical attacks.

Cryptography is a vast field with numerous techniques, algorithms, and protocols designed to secure information and communication. It is a critical component of modern information security, ensuring the confidentiality, integrity, and authenticity of data. Cryptography continues to evolve as new algorithms and techniques are developed to address emerging threats and meet the security requirements of digital systems.

9.1 RSA Encryption Algorithm

The RSA (Rivest-Shamir-Adleman) encryption algorithm is one of the most widely used public-key encryption schemes. It is named after its inventors, Ron Rivest, Adi Shamir, and Leonard Adleman, who introduced it in 1977. The RSA algorithm relies on the computational difficulty of factoring large composite numbers to achieve its security. Here's an overview of how the RSA encryption algorithm works:

Key Generation:

a. Choose two distinct prime numbers, p and q.

b. Compute their product n = p * q, which serves as the modulus for the algorithm.

c. Calculate the totient of n, $\varphi(n) = (p - 1) * (q - 1)$.

d. Select an encryption exponent e such that $1 < e < \varphi(n)$ and gcd(e, $\varphi(n)$) = 1 (e is coprime to $\varphi(n)$).

e. Compute the decryption exponent d, which is the modular multiplicative inverse of e modulo $\varphi(n)$. In other words, d is chosen such that $(d * e) \equiv 1 \pmod{\varphi(n)}$.

Encryption:

To encrypt a plaintext message M:

Number Theory

a. Represent the plaintext message as a number m, such that $0 \leq m < n$.

b. Compute the ciphertext C by raising m to the power of e and taking the result modulo n: $C \equiv m^e \pmod{n}$.

c. C is the encrypted message, the ciphertext.

Decryption:

To decrypt the ciphertext C:

a. Compute the plaintext message M by raising C to the power of d and taking the result modulo n: $M \equiv C^d \pmod{n}$.

b. M is the decrypted message, the original plaintext.

Key Security:

The security of RSA relies on the difficulty of factoring large composite numbers. To ensure strong security, the prime factors p and q used to generate the keys should be large and kept secret. The larger the prime factors and the modulus, the more secure the RSA encryption becomes.

Applications:

The RSA encryption algorithm is widely used in various applications, including:

- ✓ Secure communication: RSA enables secure transmission of information over an insecure network by encrypting messages with the recipient's public key.
- ✓ Digital signatures: RSA can be used to generate and verify digital signatures, ensuring the integrity and authenticity of digital documents.

✓ Key exchange: RSA can facilitate secure key exchange between parties who want to establish a shared secret key for symmetric encryption.

It's worth noting that as computing power increases and new cryptographic advancements occur, it is essential to keep RSA implementations up to date and adhere to recommended key lengths to maintain adequate security.

9.2 Diffie-Hellman Key Exchange

The Diffie-Hellman key exchange protocol is a method for securely establishing a shared secret key over an insecure communication channel. It allows two parties, traditionally named Alice and Bob, to agree on a shared secret without directly exchanging the secret key. The protocol was introduced by Whitfield Diffie and Martin Hellman in 1976 and is based on the computational difficulty of solving the discrete logarithm problem. Here's an overview of how the Diffie-Hellman key exchange works:

Key Generation:

a. Alice and Bob agree on a prime number p and a base g, which are public parameters.

b. Each party chooses a private key: Alice selects a random integer a, and Bob selects a random integer b.

c. Alice calculates her public key by computing $A = g^a \pmod{p}$ and sends it to Bob.

d. Bob calculates his public key by computing $B = g^b \pmod{p}$ and sends it to Alice.

Number Theory

Key Exchange:

a. Alice receives Bob's public key B, and Bob receives Alice's public key A.

b. Alice calculates the shared secret key as $S = B^a \pmod{p}$.

c. Bob calculates the shared secret key as $S = A^b \pmod{p}$.

d. Alice and Bob now have the same shared secret key S, which can be used for symmetric encryption.

Security:

The security of the Diffie-Hellman key exchange protocol relies on the computational difficulty of solving the discrete logarithm problem. Given p, g, and A (or B), it is computationally infeasible to determine the private key a (or b) from the public key A (or B). This ensures that an eavesdropper cannot determine the shared secret key by observing the exchange.

Man-in-the-Middle Attack:

The Diffie-Hellman key exchange protocol is susceptible to a man-in-the-middle attack, where an adversary intercepts and alters the exchanged public keys. To prevent this, additional mechanisms such as digital signatures or a trusted third party can be used to verify the authenticity of the exchanged keys.

Applications:

The Diffie-Hellman key exchange protocol is widely used in various applications, including:

- ✓ Secure communication: It enables parties to establish a shared secret key for symmetric encryption, ensuring the confidentiality of their communication.
- ✓ Key establishment for other cryptographic protocols: Diffie-Hellman is used as a key establishment mechanism for other cryptographic protocols, such as SSL/TLS, IPSec, and SSH.

It's important to note that while the Diffie-Hellman key exchange protocol is secure against eavesdroppers, it does not provide authentication or protection against active attacks. Additional mechanisms, such as digital signatures or encryption, are required to achieve these security properties.

9.3 Elliptic Curve Cryptography

Elliptic Curve Cryptography (ECC) is a public-key cryptography algorithm that is based on the mathematics of elliptic curves over finite fields. It offers strong security with relatively small key sizes compared to other public-key algorithms, making it efficient for resource-constrained devices. ECC has gained significant popularity and is widely used in various applications. Here's an overview of elliptic curve cryptography:

Elliptic Curves:

An elliptic curve is a mathematical curve defined by an equation of the form $y^2 = x^3 + ax + b$, where a and b are parameters. The curve's points, including a point at infinity, form an abelian group under a specific addition operation. The properties of this group make elliptic curves suitable for cryptographic applications.

Number Theory

Public and Private Keys:

ECC uses a pair of keys: a private key (randomly chosen secret) and a public key (derived from the private key). The private key is kept secret, while the public key can be freely shared with others.

Key Generation:

a. Select an elliptic curve and a base point G on the curve.

b. Choose a private key, a random integer d (typically in a certain range).

c. Compute the public key by multiplying the base point G by the private key: $Q = d * G$.

Key Exchange and Encryption:

ECC can be used for key exchange, similar to the Diffie-Hellman protocol, where two parties derive a shared secret key using their respective private keys and the other party's public key. The shared secret can then be used for symmetric encryption.

Digital Signatures:

ECC allows for secure digital signatures. To create a digital signature, the private key holder uses their private key to sign a message, and the public key holder verifies the signature using the corresponding public key. ECC signatures provide integrity, authentication, and non-repudiation of the signed data.

Security:

ECC's security is based on the elliptic curve discrete logarithm problem, which is computationally difficult to solve. Finding the private key from the public key requires solving this problem, which is believed to be computationally infeasible when using sufficiently large key sizes.

Key Sizes and Efficiency:

ECC offers strong security with smaller key sizes compared to other public-key algorithms, such as RSA and DSA. Smaller key sizes result in faster computations, lower computational resource requirements, and less bandwidth usage.

Applications:

ECC is widely used in various applications, including secure communication protocols (e.g., SSL/TLS), mobile devices, smart cards, Internet of Things (IoT), and digital signatures.

ECC provides a powerful and efficient cryptographic algorithm with strong security. Its ability to offer high security with shorter key lengths makes it particularly well-suited for resource-constrained environments where computational efficiency is crucial. However, it is essential to use proper parameter selection and adhere to recommended key sizes to ensure the desired level of security.

10. Number-Theoretic Functions

Number-theoretic functions are mathematical functions that operate on integers and provide insights into the properties and relationships of numbers. These functions play a crucial role in number theory and have applications in various areas of mathematics and computer science. Here are some important number-theoretic functions:

Euler's Totient Function (φ-function):

The Euler's totient function φ(n) counts the number of positive integers less than or equal to n that are coprime (relatively prime) to n. It measures the totient (or degree of freedom) of an integer. For a prime number p, φ(p) = p - 1. The totient function has several properties and is widely used in number theory, cryptography, and primality testing.

Mobius Function (μ-function):

The Mobius function μ(n) is a number-theoretic function that takes the value 1 if n is a square-free positive integer with an even number of prime factors, -1 if n is a square-free positive integer with an odd number of prime factors, and 0 if n is not square-free. The Mobius function is useful in various areas of number theory, such as studying prime numbers, Dirichlet convolution, and the Riemann zeta function.

Divisor Function (σ-function):

The divisor function σ(n) computes the sum of all positive divisors of an integer n, including 1 and n itself. It provides information about the factors and divisors of a number. The divisor function has variations, such as σ0(n) (the number of divisors of n) and σk(n) (the sum of the kth powers of the divisors of n).

Number Theory

Prime Counting Function (π-function):

The prime counting function π(x) counts the number of prime numbers less than or equal to a given real number x. It is used to study the distribution of prime numbers and estimate the number of primes in a given range. The prime counting function plays a crucial role in prime number theory, including the prime number theorem.

Legendre Symbol and Jacobi Symbol:

The Legendre symbol (a/p) and the Jacobi symbol (a/n) are number-theoretic symbols that provide information about the quadratic character of an integer with respect to a modulus. They are used in various applications, including primality testing, quadratic reciprocity, and cryptographic protocols.

Carmichael Function (λ-function):

The Carmichael function λ(n) computes the smallest positive integer k such that $a^k \equiv 1 \pmod{n}$ for all integers a coprime to n. It measures the order of elements in the multiplicative group modulo n and is related to Euler's totient function. The Carmichael function has applications in cryptography, particularly in RSA encryption and primality testing.

These are just a few examples of number-theoretic functions. There are many more functions, such as the Riemann zeta function, Dirichlet characters, Bernoulli numbers, and more, that are studied in number theory and have important applications in various fields of mathematics and computer science. Number-theoretic functions provide valuable insights into the properties of integers and their relationships, enabling deeper understanding and analysis of number theory concepts.

10.1 Euler's phi Function

Euler's totient function, also known as Euler's phi function or simply φ-function, is a number-theoretic function denoted by φ(n). It counts the number of positive integers less than or equal to n that are coprime (relatively prime) to n. In other words, φ(n) gives the count of positive integers between 1 and n that have no common factors with n other than 1. Here are some key properties and applications of Euler's phi function:

Definition:

Euler's totient function φ(n) is defined as the number of positive integers k ($1 \leq k \leq n$) that are coprime to n.

Coprime Integers:

Two positive integers a and b are coprime (or relatively prime) if their greatest common divisor (GCD) is 1. For example, φ(9) = 6 because the numbers 1, 2, 4, 5, 7, and 8 are coprime to 9.

Euler's Totient Theorem:

Euler's totient theorem states that for any positive integer n and its coprime integer a, $a^{\varphi(n)} \equiv 1 \pmod{n}$. This theorem is a special case of Fermat's little theorem.

Computation of φ(n):

The value of φ(n) can be computed using various methods, including prime factorization and the principle of inclusion-exclusion. For a prime number p, φ(p) = p - 1, as all positive integers less than p are coprime to p.

Multiplicative Property:

For two coprime positive integers a and b, φ(ab) = φ(a) * φ(b). This property allows the computation of φ(n) for composite numbers based on the φ-function values of their prime factors.

Relationship with Prime Numbers:

For a prime number p, φ(p) = p - 1, as all positive integers less than p are coprime to p. This property is useful in determining whether a number is prime or composite.

Applications:

Euler's totient function has various applications in number theory, cryptography, and primality testing. It is used in RSA encryption and decryption algorithms, where φ(n) is used to compute the private key. Additionally, φ-function is used in algorithms to generate and test primitive roots of prime numbers.

Euler's totient function is an important tool in number theory and has wide-ranging applications. It provides insights into the properties of integers, particularly in terms of their coprimality with other numbers. The computation of φ(n) is valuable in various cryptographic and number-theoretic algorithms, making Euler's phi function a fundamental concept in the field of mathematics.

10.2 Mobius Function

The Möbius function, denoted by μ(n), is a number-theoretic function that has important properties related to arithmetic and number theory. The Möbius function is defined for positive integers and takes on three possible values: μ(n) = 1 if n has an even number of distinct prime factors, μ(n) = -1 if n has an odd number of distinct prime factors, and μ(n) = 0 if n is divisible by a perfect square (i.e., it has repeated prime factors). Here are some key properties and applications of the Möbius function:

Definition:

The Möbius function $\mu(n)$ is defined as follows:

- ✓ $\mu(n) = 1$ if n is a square-free positive integer with an even number of prime factors.
- ✓ $\mu(n) = -1$ if n is a square-free positive integer with an odd number of prime factors.
- ✓ $\mu(n) = 0$ if n is divisible by a perfect square.

Arithmetic Functions:

The Möbius function is a multiplicative function, which means that if a and b are coprime positive integers, then $\mu(ab) = \mu(a) * \mu(b)$. This property allows the evaluation of $\mu(n)$ for composite numbers based on the prime factorization of n.

Inclusion-Exclusion Principle:

The Möbius function is related to the principle of inclusion-exclusion. It can be used to count or compute the cardinality of sets with certain properties.

Dirichlet Convolution:

The Möbius function is involved in the Dirichlet convolution, which is an operation on arithmetic functions. The Dirichlet convolution of two functions f and g is defined as $(f * g)(n) = \sum_{d|n} f(d) * g(n/d)$, where the sum is taken over all positive divisors d of n.

Relation with Prime Numbers:

The Möbius function is closely related to prime numbers. For a prime number p, μ(p) = -1, indicating that it has an odd number of prime factors. Additionally, the prime number theorem can be expressed in terms of the Möbius function.

Applications:

The Möbius function has various applications in number theory, combinatorics, and algorithm design. It is used in the study of prime numbers, including sieving algorithms such as the Möbius inversion formula. The Möbius function is also used in the Riemann zeta function, which is central to the Riemann Hypothesis.

The Möbius function is a fundamental concept in number theory, providing valuable insights into the properties of positive integers. Its properties and relationships with prime numbers and arithmetic functions make it a powerful tool in various areas of mathematics and algorithm design.

10.3 Riemann zeta Function

The Riemann zeta function, denoted by ζ(s), is a complex-valued function that is defined for complex numbers s with a real part greater than 1. It is named after the mathematician Bernhard Riemann, who extensively studied its properties. The Riemann zeta function has significant implications in number theory, complex analysis, and other areas of mathematics. Here are some key properties and features of the Riemann zeta function:

Definition:

The Riemann zeta function is defined by the infinite series:

$\zeta(s) = 1^{\wedge}(-s) + 2^{\wedge}(-s) + 3^{\wedge}(-s) + 4^{\wedge}(-s) + ...$

Number Theory

Convergence:

The series defining the Riemann zeta function converges for complex numbers s with a real part greater than 1. Other values of s, it is not defined by the series alone.

Analytic Continuation:

The Riemann zeta function can be extended to a larger domain using analytic continuation. It can be defined for all complex numbers s, excluding s = 1, where it has a simple pole. The analytic continuation reveals deep connections between the Riemann zeta function and the distribution of prime numbers.

Functional Equation:

The Riemann zeta function satisfies a functional equation that relates its values at s and 1 - s. This functional equation provides a symmetry property of the zeta function and enables calculations and investigations across the critical line of complex numbers.

Connection to Prime Numbers:

The Riemann zeta function is closely related to the distribution of prime numbers. It plays a crucial role in the study of the Riemann Hypothesis, which proposes that all non-trivial zeros of the zeta function lie on the critical line with a real part of 1/2. The Riemann Hypothesis has significant implications for prime number theory.

Special Values:

The Riemann zeta function takes on special values for certain arguments. For example, $\zeta(2) = \pi^2/6$, $\zeta(4) = \pi^4/90$, and $\zeta(0) = -1/2$.

Relationship with Other Functions:

The Riemann zeta function is connected to various other mathematical functions. For example, the values of the zeta function at positive even integers are related to Bernoulli numbers. The zeta function also appears in the study of the gamma function and other special functions.

Applications:

The Riemann zeta function has applications in various areas of mathematics, including number theory, complex analysis, harmonic analysis, and physics. It is used in the study of prime numbers, the distribution of primes, the Riemann Hypothesis, and the behavior of polynomials.

The Riemann zeta function is a fundamental and intriguing function that links together many areas of mathematics. Its properties and connections with prime numbers make it a subject of intense research and study. The zeta function's impact extends beyond number theory, providing deep insights into the behavior of complex functions and the distribution of mathematical objects.

11. Prime Number Theorems

Prime number theorems are fundamental results in number theory that provide insights into the distribution and behavior of prime numbers. These theorems help us understand the patterns and characteristics of prime numbers as we consider larger numbers. Here are two significant prime number theorems:

Prime Number Theorem:

The Prime Number Theorem, proved independently by Jacques Hadamard and Charles Jean de la Vallée Poussin in 1896, describes the asymptotic behavior of the prime-counting function $\pi(x)$, which counts the number of prime numbers less than or equal to a given real number x. The theorem states that for large values of x, the approximate number of primes up to x is given by:

$\pi(x) \sim (x / \log(x))$

Here, the symbol "~" indicates asymptotic equivalence.

This theorem implies that as x becomes larger, the density of prime numbers among the positive integers decreases. However, the ratio of $\pi(x)$ to $(x / \log(x))$ approaches 1 as x tends to infinity, indicating that the prime numbers become more evenly distributed among the positive integers.

Prime Number Theorem with Error Term:

The Prime Number Theorem also provides an error term that quantifies the discrepancy between the exact count of primes and the approximation given by $(x / \log(x))$. The error term is denoted by $O(x^{1/2} * \log(x))$, indicating that the deviation between the true count of primes and the approximation is bounded by a constant multiple of $x^{1/2} * \log(x)$ for large x.

This error term signifies that while the Prime Number Theorem gives a good approximation for the count of primes, there are small fluctuations around the approximation due to the irregularity in the distribution of prime numbers.

These prime number theorems have profound implications in number theory, the study of prime numbers, and related fields. They provide a deeper understanding of the distribution and behavior of prime numbers, allowing mathematicians to explore patterns, conjectures, and properties related to prime numbers. The Prime Number Theorem forms the basis for many advanced results in prime number theory, such as the Riemann Hypothesis and the study of prime gaps.

11.1 Prime Number Theorem

The Prime Number Theorem is a significant result in number theory that describes the asymptotic behavior of the prime-counting function $\pi(x)$, which counts the number of prime numbers less than or equal to a given real number x. The theorem was first proved independently by Jacques Hadamard and Charles Jean de la Vallée Poussin in 1896. The Prime Number Theorem states that as x becomes large, the approximate number of primes up to x is given by:

$\pi(x) \sim (x / \log(x))$

In this approximation, the symbol "~" indicates asymptotic equivalence. The Prime Number Theorem reveals that the density of prime numbers among the positive integers decreases as x increases. However, the ratio of $\pi(x)$ to $(x / \log(x))$ approaches 1 as x tends to infinity, indicating that the prime numbers become more evenly distributed among the positive integers.

The Prime Number Theorem is significant in understanding the distribution and behavior of prime numbers. It provides insights into the relationship between the size of a given interval and the number of primes within that interval. This theorem forms the foundation for many advanced results and conjectures in prime number theory, such as the study of prime gaps, the distribution of prime numbers along arithmetic progressions, and the Riemann Hypothesis.

It is important to note that while the Prime Number Theorem provides a good approximation for the count of primes, there are fluctuations around the approximation. These fluctuations are quantified by an error term, which is typically denoted by $O(x^{1/2} * \log(x))$. The error term signifies that the deviation between the true count of primes and the approximation is bounded by a constant multiple of $x^{1/2} * \log(x)$ for large x.

The Prime Number Theorem has had a profound impact on number theory and related fields. It has led to further investigations into the properties of prime numbers and their distribution, as well as the development of new conjectures and theorems. The theorem's implications extend beyond number theory, finding applications in cryptography, computer science, and other areas where prime numbers play a crucial role.

11.2 Distribution of Prime Numbers

The distribution of prime numbers is a fundamental topic in number theory that seeks to understand how prime numbers are distributed among positive integers. While primes appear to be randomly distributed, various patterns and regularities have been discovered and studied. Here are some key aspects related to the distribution of prime numbers:

Prime Number Theorem:

As mentioned earlier, the Prime Number Theorem provides an asymptotic approximation for the number of primes up to a given number x. It states that $\pi(x) \sim (x / \log(x))$, where $\pi(x)$ represents the number of primes less than or equal to x. This theorem reveals that the density of prime numbers among the positive integers decreases as x becomes larger.

Prime Number Gaps:

Prime number gaps refer to the differences between consecutive primes. While prime numbers become less frequent as numbers increase, there is no known upper bound on the gap between two consecutive primes. The study of prime gaps is a subject of ongoing research, and questions related to the existence of infinitely large prime gaps remain open.

Twin Primes and Prime Constellations:

Twin primes are pairs of prime numbers that differ by 2, such as (3, 5), (11, 13), (17, 19), and so on. Twin prime conjecture suggests that there are infinitely many twin primes. Prime constellations are groups of prime numbers that exhibit certain patterns, such as prime triplets, prime quadruplets, and prime k-tuples. The distribution and existence of such prime constellations are areas of active study.

Sieve Methods:

Sieve methods are techniques used to identify and analyze patterns in the distribution of prime numbers. The Sieve of Eratosthenes, for example, is a classic method for generating primes by iteratively crossing out multiples of primes. More advanced sieve methods, such as the Sieve of Atkin and the Sieve of Sundaram, have been developed to efficiently find primes and explore their distribution.

Prime Number Races:

Prime number races refer to the competition between certain arithmetic progressions and prime numbers. For example, the infinitude of primes in the form 4n + 1 versus the form 4n - 1 has been extensively studied. Dirichlet's theorem on arithmetic progressions establishes that for any pair of positive coprime integers a and d, there are infinitely many primes of the form a + kd, where k is a non-negative integer.

Riemann Hypothesis:

The Riemann Hypothesis is one of the most famous unsolved problems in mathematics and is closely related to the distribution of prime numbers. It proposes that all non-trivial zeros of the Riemann zeta function lie on the critical line with a real part of 1/2. The Riemann Hypothesis, if proven true, would provide deep insights into the distribution of prime numbers.

The distribution of prime numbers continues to be an active area of research, with mathematicians striving to uncover further patterns, establish conjectures, and prove theorems. While many questions remain unanswered, the study of prime numbers and their distribution plays a fundamental role in number theory and has applications in cryptography, prime factorization algorithms, and various areas of mathematics and computer science.

11.3 Riemann Hypothesis

The Riemann Hypothesis is one of the most famous unsolved problems in mathematics, named after the mathematician Bernhard Riemann who proposed it in 1859. The hypothesis is closely related to

Number Theory

the distribution of prime numbers and the behavior of the Riemann zeta function.

The Riemann Hypothesis states that all non-trivial zeros of the Riemann zeta function, denoted by $\zeta(s)$, lie on the critical line with a real part of 1/2. The Riemann zeta function is a complex-valued function defined for complex numbers s with a real part greater than 1. It is expressed as the infinite series $\zeta(s) = 1^{-s} + 2^{-s} + 3^{-s} + 4^{-s} + ...$

Here are some key aspects of the Riemann Hypothesis:

Critical Line:

The critical line is the vertical line in the complex plane defined by $Re(s) = 1/2$, where $Re(s)$ represents the real part of s. The Riemann Hypothesis states that all non-trivial zeros of the zeta function lie on this critical line. Trivial zeros are negative even integers, such as -2, -4, -6, and so on.

Connection to Prime Numbers:

The Riemann Hypothesis has profound implications for the distribution of prime numbers. It provides precise information about the spacing of prime numbers and the occurrence of prime number patterns. If the hypothesis is true, it would establish a deep connection between the distribution of primes and the behavior of the Riemann zeta function.

Significance:

The Riemann Hypothesis is considered one of the most important unsolved problems in mathematics. Its resolution would have far-reaching consequences in number theory and related fields. It would provide deeper insights into prime number theory, improve our understanding of the distribution of primes, and lead to advancements in related mathematical topics.

Implications for Prime Number Theorems:

The Riemann Hypothesis is closely linked to prime number theorems, such as the Prime Number Theorem and the distribution of prime gaps. If the hypothesis is proven true, it would yield sharper bounds and precise information about the distribution of prime numbers, confirming various conjectures and advancing our understanding of prime number theory.

Computational Verification:

While extensive numerical evidence supports the Riemann Hypothesis, a general proof or disproof has not yet been found. Researchers have verified the hypothesis for an extensive range of non-trivial zeros using computational methods, but complete analytical proof remains elusive.

Efforts to prove or disprove the Riemann Hypothesis have attracted the attention of many mathematicians over the years, and it continues to be an active area of research. Progress has been made in understanding the behavior of the zeta function and its zeros, but a definitive resolution of the Riemann Hypothesis remains an open problem that captivates the mathematical community.

12. Unsolved Problems in Number Theory

Number theory, as a branch of mathematics, is replete with fascinating problems that remain unsolved despite decades or even centuries of efforts by mathematicians. Here are some notable unsolved problems in number theory:

Riemann Hypothesis:

The Riemann Hypothesis, mentioned earlier, is one of the most famous unsolved problems in mathematics. It relates to the distribution of prime numbers and the behavior of the Riemann zeta function, proposing that all non-trivial zeros of the zeta function lie on the critical line with a real part of 1/2.

Twin Prime Conjecture:

The Twin Prime Conjecture states that there are infinitely many pairs of prime numbers that differ by 2, such as (3, 5), (11, 13), (17, 19), and so on. Although twin primes have been observed up to very large values, proving that infinitely many exist remains an open problem.

Goldbach's Conjecture:

Goldbach's Conjecture states that every even integer greater than 2 can be expressed as the sum of two prime numbers. For example, $10 = 3 + 7$ and $24 = 11 + 13$. While the conjecture has been tested extensively for even numbers up to enormous values, a general proof or disproof remains elusive.

Collatz Conjecture:

The Collatz Conjecture, also known as the $3n + 1$ problem, proposes that for any positive integer n, the sequence obtained by repeatedly applying the transformation n → n/2 (if n is even) or n → 3n + 1 (if n is odd) will eventually reach the number 1. Despite extensive computational verification, a general proof for all positive integers is yet to be found.

Beal's Conjecture:

Beal's Conjecture, posed by Andrew Beal in 1993, suggests that if A, B, and C are positive integers, and $A^x + B^y = C^z$ for positive integers x, y, and z with x, y, z > 2, then A, B, and C must have a common prime factor. This conjecture remains unproven, and its resolution has connections to Fermat's Last Theorem and other related problems.

ABC Conjecture:

The ABC Conjecture, formulated by Joseph Oesterlé and David Masser in 1985, involves the relationship between the prime factors of three positive integers A, B, and C, such that A + B = C. The conjecture proposes that if A, B, and C have no common prime factors and C is large enough, then the product of the distinct prime factors of ABC is typically much smaller than C. Although progress has been made, a general proof or disproof remains unknown.

These are just a few examples of unsolved problems in number theory. Number theory is a rich and fertile field with numerous intriguing problems waiting to be explored and solved. While mathematicians continue their quest for solutions, these unsolved problems contribute to ongoing research, spark new discoveries, and inspire further investigations into the deep mysteries of number theory.

12.1 Goldbach's Conjecture

Goldbach's Conjecture is one of the oldest and most famous unsolved problems in number theory. Proposed by the German mathematician Christian Goldbach in 1742, the conjecture states that every even integer greater than 2 can be expressed as the sum of two prime numbers. Formally, it can be stated as follows:

"For every even integer $n > 2$, there exist prime numbers p and q such that $n = p + q$."

Despite being extensively tested for even numbers up to enormous values, a general proof or disproof of Goldbach's Conjecture remains elusive. However, many numerical verifications and partial results have been obtained over the years, reinforcing the belief that the conjecture is likely true.

Efforts to prove Goldbach's Conjecture have involved various strategies, including advanced mathematical techniques and computer-assisted calculations. Significant progress has been made in establishing weaker versions of the conjecture, such as the "ternary Goldbach conjecture," which allows for the representation of even numbers as the sum of three prime numbers.

Historically, Goldbach's Conjecture has gained attention due to its simplicity and the tantalizing possibility that every even integer can be expressed as a sum of primes. Numerous mathematicians, including famous figures like Leonhard Euler, have contributed to the study of this conjecture. Despite their efforts, a complete proof or counterexample has yet to be found.

Goldbach's Conjecture continues to inspire and challenge mathematicians, and its resolution would have profound implications for number theory and the understanding of prime numbers. The search for a proof or disproof of this conjecture remains an active area of research, with mathematicians employing a range of techniques and insights to unlock the mystery of Goldbach's Conjecture.

12.2 Twin Prime Conjecture

The Twin Prime Conjecture is a famous unsolved problem in number theory. It states that there are infinitely many pairs of prime numbers that differ by 2. These pairs of primes are called "twin primes." For example, (3, 5), (11, 13), (17, 19), and so on.

While twin primes have been observed up to very large values, proving that infinitely many twin primes exist remains an open problem. This conjecture has a long history and has been studied by numerous mathematicians, including Pierre de Fermat and Christian Goldbach.

Efforts to prove the Twin Prime Conjecture have led to significant progress in the study of prime numbers and related areas. In 2013, Yitang Zhang made a breakthrough by proving that there exist infinitely many pairs of primes that differ by a finite distance (specifically, within 70 million). Although this result didn't establish the infinitude of twin primes directly, it demonstrated the plausibility of the conjecture.

Later, Zhang's work was improved upon by subsequent mathematicians, leading to even smaller bounds on the prime gaps. This progress has increased confidence in the likelihood of infinitely many twin primes, but a complete proof for all pairs of primes still eludes mathematicians.

Researchers continue to investigate the Twin Prime Conjecture, searching for new techniques and insights that might lead to a proof. The conjecture is connected to other important questions in prime number theory, such as the distribution of prime numbers along arithmetic progressions.

The Twin Prime Conjecture captures the fascination of mathematicians and the general public alike. Its resolution would contribute to a deeper understanding of the distribution of prime numbers and reveal intricate patterns within this fundamental aspect of number theory. Despite the ongoing efforts, the Twin Prime Conjecture remains an open problem, awaiting a breakthrough that could bring us closer to a definitive answer.

12.3 Collatz Conjecture

The Collatz Conjecture, also known as the 3n + 1 problem or the Syracuse problem, is an unsolved problem in mathematics. It was proposed by German mathematician Lothar Collatz in 1937. The conjecture is a simple iterative process applied to positive integers.

The conjecture can be described as follows:

1. Start with any positive integer n.

2. If n is even, divide it by 2 (n/2).

3. If n is odd, multiply it by 3 and add 1 (3n + 1).

4. Repeat the process with the resulting number.

The Collatz Conjecture states that no matter what positive integer is chosen as the starting point, the sequence eventually reaches the number 1 and then cycles between 1 and 4 indefinitely. In other words, for any positive integer n, the sequence will eventually reach the loop $1 \to 4 \to 2 \to 1$.

Despite extensive computational verification for an enormous range of starting values, a general proof for all positive integers remains elusive. The Collatz Conjecture presents a unique challenge due to the unpredictable nature of the sequence and the intricate relationships between even and odd numbers.

Efforts to prove the Collatz Conjecture have involved various mathematical techniques, including number theory, graph theory, and computer-assisted calculations. While progress has been made in understanding certain properties and patterns within the sequence, a comprehensive proof or disproof remains an open problem.

The Collatz Conjecture has captured the interest of mathematicians, and many researchers continue to explore this intriguing problem. Resolving the conjecture would not only contribute to our understanding of number theory but also shed light on the behavior of iterative processes and the relationship between even and odd numbers.

Despite decades of investigation, the Collatz Conjecture continues to resist complete resolution, leaving mathematicians inspired by its complexity and searching for new insights that might unravel its mysteries.

13. Conclusion and Further Exploration

In conclusion, number theory is a fascinating branch of mathematics that deals with the properties, relationships, and patterns of integers and their subsets, particularly prime numbers. It has significant applications in cryptography, algorithm design, and other areas of mathematics.

Throughout this discussion, we explored various topics in number theory, including divisibility and factors, prime numbers, modular arithmetic, Diophantine equations, number-theoretic functions, quadratic residues, continued fractions, and cryptography. We also touched upon some famous unsolved problems like the Riemann Hypothesis, Goldbach's Conjecture, Twin Prime Conjecture, and Collatz Conjecture.

However, this is just the tip of the iceberg when it comes to number theory. There are many other fascinating topics and unsolved problems within this field that offer opportunities for further exploration. Some of these include:

- ✓ Prime gaps: Investigating the distribution and characteristics of prime gaps, such as the occurrence of consecutive primes with large gaps or small gaps.
- ✓ Arithmetic progressions: Studying prime numbers occurring in arithmetic progressions, including questions related to the infinitude of primes in certain arithmetic progressions.
- ✓ Diophantine approximation: Exploring equations and inequalities with integer solutions, such as finding rational approximations to irrational numbers.
- ✓ Primality testing: Developing efficient algorithms to determine whether a given number is prime.

- ✓ Analytic number theory: Applying techniques from complex analysis and calculus to study number-theoretic functions and their properties.
- ✓ Algebraic number theory: Investigating number systems beyond the rational numbers, such as algebraic integers and algebraic extensions of rational numbers.

These topics provide avenues for further exploration and research within number theory. They offer opportunities to deepen our understanding of the intricate world of numbers and uncover new insights and discoveries.

Whether you're a mathematician, a student, or simply someone with a curious mind, delving deeper into number theory can be a rewarding endeavor. The field offers a rich blend of theory, problem-solving, and mathematical elegance. So, go ahead and continue your exploration of number theory, and who knows, you may even contribute to the solution of one of its unsolved problems!

www.ingramcontent.com/pod-product-compliance
Lightning Source LLC
Chambersburg PA
CBHW072225170526
45158CB00002BA/758